Springer Series in Solid-State Sciences

Volume 186

The Springer Series in Solid-State Sciences consists of fundamental scientific books prepared by leading researchers in the field. They strive to communicate, in a systematic and comprehensive way, the basic principles as well as new developments in theoretical and experimental solid-state physics.

More information about this series at http://www.springer.com/series/682

Ferdinando Mancini · Roberta Citro
Editors

The Iron Pnictide Superconductors

An Introduction and Overview

Springer

Editors
Ferdinando Mancini
International Institute for Advance Scientific Studies
Vietri sul Mare, Salerno
Italy

Roberta Citro
Dipartimento di Fisica "E.R. Caianiello"
Università degli Studi di Salerno
Fisciano, Salerno
Italy

ISSN 0171-1873 ISSN 2197-4179 (electronic)
Springer Series in Solid-State Sciences
ISBN 978-3-319-56116-5 ISBN 978-3-319-56117-2 (eBook)
DOI 10.1007/978-3-319-56117-2

Library of Congress Control Number: 2017936466

Printed on acid-free paper

This Springer imprint is published by Springer Nature
The registered company is Springer International Publishing AG
The registered company address is: Gewerbestrasse 11, 6330 Cham, Switzerland

Preface

This volume “The Iron Pnictide Superconductors—An Introduction and Overview” covers different theoretical aspects of the physics of iron-based superconductors ranging from the analytical to numerical and computational ones. It starts from the basic theory modeling many-body physics in Fe-superconductors and other multi-orbital materials to go to the studies on magnetic, superconducting, and nematic order instabilities. Finally, it offers a comprehensive overview of the recent advancements in the theoretical investigations of Mott transition and strong correlations in iron-based superconductors.

The book contains the notes of the lectures delivered at the “Nineteenth Training Course in the Physics of Strongly Correlated Systems” held in Vietri sul Mare (Salerno, Italy) in October 2015. The course consisted of four lectures every morning, held by Professors Bernd Büchner, Massimo Capone, Luca de’ Medici, Ilya Eremin, and JörgSchmalian, and afternoon activities (seminars delivered by the junior researchers, solving of specific problems, roundtable on hot topics, hands-on training on relevant numerical issues…) aimed principally at promoting discussions between the attendees and the lecturers. The outcome of this type of course was a significant interchange of ideas among the participants thanks to both the enlightening morning lectures and the long afternoon sessions devoted to discussions.

The book has both introductory and pedagogical aspects that could be very useful for researchers entering the field of unconventional superconductivity. The readers will strongly benefit from the different overviews on the topic of multi-band superconductivity in iron-based superconductors.

We wish to acknowledge the support of those institutions that made the course possible. The main sponsors of the event were the Department of Physics “E.R. Caianiello”—University of Salerno and the International Institute for Advanced Scientific Studies “E.R. Caianiello.”

Salerno, Italy

Ferdinando Mancini
Roberta Citro

Contents

Contributors

Felix Ahn Institut Für Theoretische Festkörperphysik III, Ruhr-Universität Bochum, Bochum, Germany

Massimo Capone SISSA, Trieste, Italy

Roberta Citro Dipartimento di Fisica “E.R. Caianiello”, Università degli Studi di Salerno, Fisciano (SA), Italy; IIASS, International Institute for Advanced Scientific Studies “E.R. Caianiello”, Vietri sul Mare (SA), Italy

Luca de’ Medici ESPCI, Paris, France

Ilya Eremin Institut Für Theoretische Festkörperphysik III, Ruhr-Universität Bochum, Bochum, Germany

Rafael M. Fernandes School of Physics and Astronomy, University of Minnesota, Minneapolis, MN, USA

Una Karahasanovic Institut für Theorie der Kondensierten Materie and Institut für Festkörperphysik, Karlsruher Institut für Technologie, Karlsruhe, Germany

Johannes Knolle TCM Group, Cavendish Laboratory, Cambridge, UK

Ferdinando Mancini Dipartimento di Fisica “E.R. Caianiello”, Università degli Studi di Salerno, Fisciano (SA), Italy; IIASS, International Institute for Advanced Scientific Studies “E.R. Caianiello”, Vietri sul Mare (SA), Italy

Joerg Schmalian Institut für Theorie der Kondensierten Materie and Institut für Festkörperphysik, Karlsruher Institut für Technologie, Karlsruhe, Germany

Chapter 1
Iron Based Supercondutors: Introduction to the Volume

Roberta Citro and Ferdinando Mancini

Abstract A brief introduction on the content of the volume with a summary of the main studies in each chapter is reported. The volume covers the physics of iron based superconductors from a theoretical, numerical and computational point of view. It also contains a section dedicated to training, which could be beneficial to researchers approaching the field of unconventional superconductivity. The readers will strongly benefit from the different overviews on the topic of iron based superconductors and multi-orbital systems. It starts from the basic theory of modelling many-body physics in Fe-superconductors to go to the magnetic and Cooper pair fluctuations and nematic order. Lastly, it offers a comprehensive overview of the most recent advances on the role of interactions in iron-based superconductors.

Ever since their discovery, in 2008, *Fe*-based pnictides superconductors [18] (binary compounds of the elements from the 5th group: N, P, As, Sb, Bi) have represented a significant breakthrough in Condensed Matter Physics. A great deal of effort, by both theoretical and experimental condensed matter community, has been devoted in the last decade to the understanding of normal state properties of these materials, including magnetism, pairing mechanism, and the symmetry and structure of the pairing gap.

The family of *Fe*-based superconductors (FeSC) is already quite large: it includes various Fe-pnictides such as 1111 systems RFeAsO (R = rare earth element) [5, 6, 18, 34], 122 systems XFe_2As_2 (X = alkaline earth metals) [30, 36, 37], 111 systems like LiFeAs [41], and also Fe-chalcogenides (Fe-based compounds with

R. Citro (✉) · F. Mancini
Dipartimento di Fisica "E.R. Caianiello", Università degli Studi di Salerno, 84084 Fisciano (SA), Italy
e-mail: citro@sa.infn.it

F. Mancini
e-mail: mancini@sa.infn.it

R. Citro · F. Mancini
IIASS, International Institute for Advanced Scientific Studies "E.R. Caianiello", 84019 Vietri sul Mare (SA), Italy

F. Mancini and R. Citro (eds.), *The Iron Pnictide Superconductors*, Springer Series in Solid-State Sciences 186, DOI 10.1007/978-3-319-56117-2_1

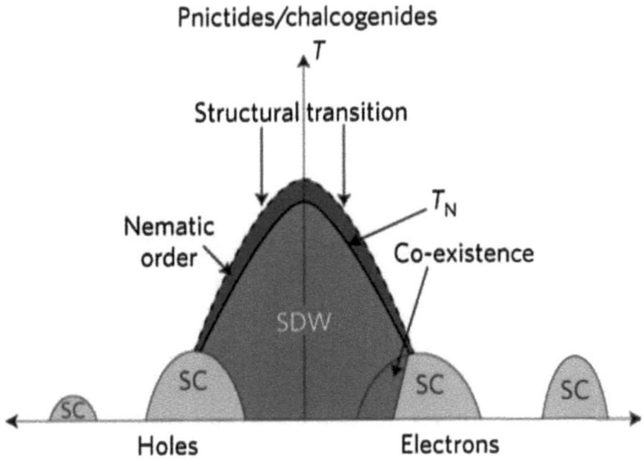

Fig. 1.1 Phase diagram of Fe-based pnictides upon hole or electron doping. In the shaded region, superconductivity and antiferromagnetism co-exist. Superconductivity can be initiated, not only by doping, but also by pressure and/or isovalent replacement of one pnictide element with another [29]. Nematic phase at $T > T_N$ is subject of a chapter of the book. Superconductors at large doping are KFe_2As_2 for hole doping and $A_xFe_{2-y}Se_2$ (A = K, Rb, Cs) for electron doping [14, 33]. Figure from [2]

elements from the 16th group: S, Se, Te) such as $FeTe_{1-x}Se_x$ [28] and AFe_sSe_2 $(A = K, Rb, Cs)$ [14, 33].

The parent compounds of Fe-based superconductors are metals, differently from cuprate superconductors whose parent compounds are Mott insulators. Moreover, as for cuprates, in most cases these parent compounds are antiferromagnetically ordered [11]. Superconductivity (SC) in FeSCs emerges on either hole or electron doping (see Fig. 1.1), but can also be induced by pressure or by isovalent replacement of one pnictide element with another, e.g., As by P [29]. In some systems, like LiFeAs [41] and LaFePO [17], SC emerges already at zero doping, instead of a magnetic order.

This phase diagram shows strong similarities with the generic cuprate one and gives evidence for the interplay of magnetism and superconductivity in iron-based materials. The magnetism, electronic structure, normal state properties of FeSCs, and the interplay between FeSCs and cuprate superconductors have been reviewed in several recent publications [2, 7, 8, 13, 15, 16, 20, 23–26, 31, 43]. This Volume is an attempt to overview our current understanding of the magnetism and pairing mechanism, the symmetry and structure of the pairing gap at various hole and electron doping. It also includes a systematic study on the effect of interactions beyond a simple mean-field level.

The goal of Chap. 2 is two fold: on the one hand, it provides a pedagogical description on how the antiferromagnetic order is formed as a consequence of the electronic structures, orbital degrees of freedom, as well as, interactions. The role of the orbitals, which are present at the Fermi surface, in determining the momentum dependence of the superconducting and magnetic order parameters is discussed. The theoretical description is based on the unambiguous experimental observation that parent ferropnictides are metals of *itinerant electrons*. They are not Mott insulators whose magnetism would be described by localized spin models. For example, the $J_1 - J_2$ model of localized moments with nearest and next-nearest neighbour interactions displays the correct stripe-like magnetism for $J_2 > 0.5J_1$ [4, 38, 39, 44] and many researchers have advocated this description also for iron based superconductors. In contrast, here the description is based on the paradigmatic metallic magnet Cr [19, 35] where even weak interactions trigger a transition to an incommensurate antiferromagnetic state due to a special nesting-property of the band structure. Given the electronic structure of ferropnictides, it is natural to assume that antiferromagnetic order emerges, at least partly, due to near nesting between the dispersion of holes and electrons. Assuming weak to moderate interactions, an itinerant spin density wave scenario for iron based superconductors is advocated [1, 3, 9, 21, 22, 32, 42].

In the second part of this chapter the way to calculate the spin excitations in the magnetically ordered state for the itinerant magnetism is described. In contrast to localized spin models, the task is more demanding because dealing with spin and charge degrees of freedom simultaneously.

As first, the formalism for the single-band Hubbard model in the simple (π, π) antiferromagnetic state is reviewed, followed by a discussion on the situation of multi-band ferropnictides with stripe-like magnetism. The presence of additional orbital degrees of freedom makes the theory more involved thus arising new interesting phenomena. The competing nesting vectors at $(\pi, 0)$ and $(0, \pi)$ provide a source of a new kind of *frustration* which is normally only studied in insulating magnets. The details of the competition between antiferromagnetism and superconductivity in iron-based superconductors are not discussed, as this subject is already reported elsewhere [8].

The aim of Chap. 3 is to provide a semi-phenomenological explanation of the phase diagram of iron based superconductors, and summarize theoretical predictions for the signatures of the nematic degrees of freedom in various experimentally measurable quantities. In most cases the study will be based on a spin-driven nematic approach and the focus will be on the:

- construction of a phase diagram for iron-based superconductors;
- development of a φ^4 theory of the nematic phase and the prediction on the nature of the nematic transition;
- predictions on the behavior of the elastic modulus in the vicinity of a nematic/structural transition;
- determination of the Raman response function, in both the normal and superconducting state;
- determination of the resistivity anisotropy;
- discussion on the signatures of the Ising-nematic state in the magnetic spectrum.

Lastly, the impact of nematic fluctuations on superconductivity will be briefly discussed.

Signatures of the electron nematic phases have recently been observed in $2d$ electron gases in magnetic fields [12] and in the vicinity of striped phases of cuprates (for a review see for example [40]) and pnictides (for a review see [10]). They show anisotropies in transport measurements which can be detected in STM measurements, ARPES, neutron scattering experiments and NMR, as well as trigger simultaneous structural transitions and orbital ordering in pnictides. The focus of the chapter is on the so-called Ising-nematic state in iron-based superconductors, which occurs above the striped spin-density wave phase and tracks the magnetic transition closely. The main feature of this phase is the lowering of the tetragonal symmetry to orthorhombic, while preserving spin-rotational invariance. Several theoretical proposals have been developed to explain its existence, mostly based on two different scenarios. The former proposes orbital ordering and fluctuations as a driving force behind the nematic transition, while in the latter, spin fluctuations associated with striped magnetic order generate the emergent electronic nematic order at temperatures above the Néel temperature. The primary argument for the spin-driven nematic approach is the vicinity or coincidence of the nematic and magnetic transition temperatures in the phase diagram.

The goal of Chap. 4 is to expose the slave-spin mean-field theory, an accurate technique to model electronic correlations in Fermi-liquid phases of multi-orbital materials, with a focus on its recent successful application to the physics of iron-based superconductors. Apart from introducing electronic correlations and recalling the Fermi-liquid phenomenology, the chapter offers a step-by-step explanation of the slave-spin technique with a set of useful complements providing an analytical insight into Mott and Hund's physics, which are at the core of the Physics of strongly correlated materials. Some original research material is also exposed, such as the Hund-induced shrinking of the Hubbard bands flanking the gap of the half-filled Mott insulator, and the low-energy description of the "orbital-decoupling" mechanism.

The origin of the study presented in this chapter is the above mentioned chemical and structural similarities between cuprates and iron-based materials which have led, at least a part of the community, to pursue the analogy also in more microscopic terms, therefore proposing that electronic correlations have to play a central role also in the iron-based material. Nevertheless, this point of view has been immediately challenged by the lack of Mott insulators in the iron-based family, which casts doubts on the strength of the electronic correlations and their possible role in driving or affecting the superconducting phase. This, and many other seemingly conflicting evidences, have led the community to split up: some adopting a weak-coupling approach and others a strong-coupling one. The former is corroborated by the success of the Density Functional Theory (DFT) to reproduce many aspects of the band-structure and the Fermi surface, which is typically a fingerprint of minor correlation effects. Within this approach magnetism is due to Fermi-surface nesting of otherwise metallic mobile carriers and it has an itinerant character, while superconductivity arises due to the exchange of bosons of magnetic nature [27]. Instead, the latter assumes that the iron-based materials are close to Mott localization and describes magnetism as the ordering of almost localized magnetic moments. In this view, the carriers are poorly mobile and the main interactions can be modelled in terms of a frustrated Heisenberg model with nearest and next-nearest-neighbor superexchange interactions [38]. In the final part of this chapter it is shown that a sort of intermediate scenario, where both localized and delocalized carriers are simultaneously present, can overcome this dualism and account for most of the properties of the iron-based phase diagram.

References

1. A. Abanov, A.V. Chubukov, M.R. Norman, Phys. Rev. B **78**, 220507(R) (2008)
2. D.N. Basov, A.V. Chubukov, Nat. Phys. **7**, 241 (2011)
3. P.M.R. Brydon, C. Timm, Phys. Rev. B **80**, 174401 (2009); **79**, 180504(R) (2009)
4. P. Chandra, P. Coleman, A.I. Larkin, Phys. Rev. Lett. **64**, 88 (1990)
5. G.F. Chen, Z. Li, D. Wu, G. Li, W.Z. Hu et al., Phys. Rev. Lett. **100**, 247002 (2008)
6. X.H. Chen, T. Wu, G. Wu, R.H. Liu, H. Chen, D.F. Fang, Nature **453**, 761 (2008)
7. A.V. Chubukov, Phys. C **469**, 640 (2009)
8. A.V. Chubukov, Annu. Rev. Condens. Matter Phys. **3**, 13 (2012)
9. V. Cvetkovic, Z. Tesanovic, Europhys. Lett. **85**, 37002 (2009)
10. R.M. Fernandes, J. Schmalian, Supercond. Sci. Technol. **25**, 084005 (2012)
11. For the latest results on magnetic measaurements, see D.S. Inosov, J.T. Park, P. Bourges, D.L. Sun, Y. Sidis, A. Schneidewind, K. Hradil, D. Haug, C.T. Lin, B. Keimer, V. Hinkov, Nat. Phys. **6**, 178–181 (2010) and references therein
12. E. Fradkin, S.A. Kivelson, E. Manousakis, K. Nho, Phys. Rev. Lett. **84**, 1982–1985 (2000)
13. S. Graser, T.A. Maier, P.J. Hirshfeld, D.J. Scalapino, New J. Phys. **11**, 025016 (2009)
14. J. Guo, S. Jin, G. Wang, S. Wang, K. Zhu, T. Zhou, M. He, X. Chen, Phys. Rev. B **82**, 180520(R) (2010)
15. P.J. Hirschfeld, M.M. Korshunov, I.I. Mazin, arXiv:1106.3712
16. D.C. Johnston, Adv. Phys. **59**, 803 (2010)
17. Y. Kamihara, T. Watanabe, M. Hirano, H. Hosono, J. Am. Chem. Soc. **128**, 10012 (2006)
18. Y. Kamihara, T. Watanabe, M. Hirano, H. Hosono, J. Am. Chem. Soc. **130**, 3296 (2008)
19. L.V. Keldysh, Yu.V. Kopaev, Sov. Phys. Solid State **6**, 2219 (1965)
20. A.F. Kemper, T.A. Maier, S. Graser, H.-P. Cheng, P.J. Hirschfeld, D.J. Scalapino, New J. Phys. **12**, 073030 (2010)
21. J. Knolle, I. Eremin, A. Akbari, R. Moessner, Phys. Rev. Lett. **104**, 257001 (2010)
22. J. Knolle, I. Eremin, A.V. Chubukov, R. Moessner, Phys. Rev. B **81**, 140506 (2010)
23. K. Kuroki, H. Usui, S. Onari, R. Arita, H. Aoki, Phys. Rev. B **79**, 224511 (2009)
24. S. Maiti, M.M. Korshunov, T.A. Maier, P.J. Hirschfeld, A.V. Chubukov, Phys. Rev. B **84**, 224505 (2011)
25. I.I. Mazin, Nature **464**, 183 (2010)
26. I.I. Mazin, J. Schmalian, Phys. C **469**, 614 (2009)
27. I. Mazin, D.J. Singh, M.D. Johannes, M.H. Du, Phys. Rev. Lett. **101**, 076401 (2008)
28. Y. Mizuguchi, F. Tomioka, S. Tsuda, T. Yamaguchi, Y. Takano, Appl. Phys. Lett. **93**, 152505 (2008); F.C. Hsu et al., Proc. Natl. Acad. Sci. U.S.A. **105**, 14262 (2008), M.H. Fang, H.M. Pham, B. Qian, T.J. Liu, E.K. Vehstedt, Y. Liu, L. Spinu, Z.Q. Mao, Phys. Rev. B **78**, 224503 (2008), G.F. Chen, Z.G. Chen, J. Dong, W.Z. Hu, G. Li, X.D. Zhang, P. Zheng, J.L. Luo, N.L. Wang, Phys. Rev. B **79**, 140509(R) (2009)
29. Y. Nakai, T. Iye, S. Kitagawa, K. Ishida, H. Ikeda, S. Kasahara, H. Shishido, T. Shibauchi, Y. Matsuda, T. Terashima, Phys. Rev. Lett. **105**, 107003 (2010)
30. N. Ni, A. Thaler, J.Q. Yan, A. Kracher, E. Colombier, S.L. Bud'ko, P.C. Canfield, Phys. Rev. B **82**, 024519 (2010)
31. J.-P. Paglione, R.L. Greene, Nat. Phys. **6**, 645 (2010)
32. C. Platt, C. Honerkamp, W. Hanke, New J. Phys. **11**, 055058 (2009)
33. T. Qian, X.-P. Wang, W.-C. Jin, P. Zhang, P. Richard, G. Xu, X. Dai, Z. Fang, J.-G. Guo, X.-L. Chen, H. Ding, Phys. Rev. Lett. **106**, 187001 (2011)

34. Z.-A. Ren, G.-C. Che, X.-L. Dong, J. Yang, W. Lu et al., Europhys. Lett. **83**, 17002 (2008)
35. M.T. Rice, Phys. Rev. B **2**, 3619 (1970)
36. M. Rotter, M. Tegel, D. Johrendt, Phys. Rev. Lett. **101**, 107006 (2008)
37. K. Sasmal, B. Lv, B. Lorenz, A.M. Guloy, F. Chen, Y.-Y. Xue, C.-W. Chu, Phys. Rev. Lett. **101**, 107007 (2008)
38. Q. Si, E. Abrahams, Phys. Rev. Lett. **101**, 076401 (2008)
39. G.S. Uhrig, M. Holt, J. Oitmaa, O.P. Sushkov, R.R.P. Singh, Phys. Rev. B **79**, 092416 (2009)
40. M. Vojta, Adv. Phys. **58**, 699820 (2009)
41. X.C. Wang, Q.Q. Liu, Y.X. Lv, W.B. Gao, L.X. Yang, R.C. Yu, F.Y. Li, C.Q. Jin, arXiv:0806.4688v3; S.V. Borisenko, V.B. Zabolotnyy, D.V. Evtushinsky, T.K. Kim, I.V. Morozov et al., Phys. Rev. Lett. **105**, 067002 (2010)
42. F. Wang, H. Zhai, Y. Ran, A. Vishwanath, D.-H. Lee, Phys. Rev. Lett. **102**, 047005 (2009)
43. H.H. Wen, S. Li, Annu. Rev. Condens. Matter Phys. **2**, 121 (2011)
44. C. Xu, M. Muller, S. Sachdev, Phys. Rev. B **78**, 020501(R) (2008)

Chapter 2
Itinerant Magnetic Order and Multiorbital Effects in Iron-Based Superconductors

Felix Ahn, Johannes Knolle and Ilya Eremin

2.1 Introduction

A comprehensive investigation of the relationship between magnetism and superconductivity in the iron-based superconductors, discovered in 2008 by Hideo Hosono and collaborators [30], yielded significant advances in our understanding of the magnetic ground states in these compounds and interaction with superconductivity. It is believed that the same magnetic interactions that drive the magnetic ordering also produce the Cooper-pairing [29]. The phase diagram of ferropnictides (FPs) is similar to high-T_c cuprates and contains an antiferromagnetic (AF) phase in close proximity to the superconducting (SC) one. Most iron-based superconductors exhibit an antiferromagnetic state at low carrier concentrations which is suppressed with doping, pressure, or disorder allowing for the emergence of superconductivity. This shows strong similarities to the generic cuprate phase diagram and is evidence for the interplay of magnetism and superconductivity in the iron-based materials.

The electronic structure of parent ferropnictides in the normal state has been measured by angle-resolved photoemission (ARPES) [17, 43, 44, 69, 77, 81] and by magneto-oscillations [4, 12]. Both agree largely with ab-initio band structure calculations [52, 65]. It consists of two quasi-two-dimensional near-circular hole pockets of unequal size, centered around the Γ-point (0, 0), and two quasi-2D elliptic electron pockets centered around $(0, \pi)$ and $(\pi, 0)$ points in the unfolded Brillouin

F. Ahn · I. Eremin (✉)
Institut Für Theoretische Festkörperphysik III, Ruhr-Universität Bochum,
44801 Bochum, Germany
e-mail: ilya.eremin@rub.de

F. Ahn
e-mail: felix.ahn@rub.de

J. Knolle
TCM Group, Cavendish Laboratory, 19 JJ Thomson Avenue, Cambridge 3 0HE, UK
e-mail: jk628@cam.ac.uk

F. Mancini and R. Citro (eds.), *The Iron Pnictide Superconductors*,
Springer Series in Solid-State Sciences 186, DOI 10.1007/978-3-319-56117-2_2

zone (BZ) which includes only Fe atoms. For tetragonal symmetry, the two electron pockets transform into each other under rotation by 90°. In the folded Brillouin zone, which is used for experimental measurements because of two nonequivalent As positions with respect to an Fe plane, both electron pockets are centered around (π, π). The dispersions near electron pockets and near hole pockets are reasonably close to each other apart from the sign change, i.e., there is a substantial degree of nesting between hole and electron bands.

In the following notes we discuss the peculiarity of the magnetism observed as the parent phase of most of the iron based superconductors from the itinerant point of view. The magnetic order is quasi two dimensional due to the layered structure of the materials and consists of ferromagnetic (FM) chains along one crystallographic direction which alternate antiferromagnetically along the other direction. In reciprocal space this corresponds to an ordering momentum of either $(0, \pi)$, or $(\pi, 0)$ (in the Fe-only Brillouin zone). This order breaks not only spin rotational and translational symmetry but also the rotational symmetry of the lattice because the FM chains can point in one or the other direction. It is the latter additional Z_2 symmetry which makes magnetism more interesting in iron based materials in contrast to the simple (π, π) AFM of the cuprate high-temperature superconductors. In addition, it gives rise to a host of new phenomena a theoretical understanding of which is crucial also for a uncovering the mechanism of superconductivity.

In this chapter, our goal is two fold: on the one hand, we aim to give a pedagogical description how the antiferromagnetic order forms as a consequence the electronic structures, orbital degrees of freedom, as well as, interactions. We discuss the role of the orbitals, which are present at the Fermi surface, in determining the momentum dependence of the superconducting and magnetic order parameters. Our theoretical description is based on the unambiguous experimental observation that parent ferropnictides are metals of *itinerant electrons*. They are not Mott insulators whose magnetism would be described by localized spin models. For example, the $J_1 - J_2$ model of localized moments with nearest and next-nearest neighbour interactions displays the correct stripe like magnetism for $J_2 > 0.5J_1$ [5, 64, 71, 75, 79] and various people have advocated this description also for iron based superconductors. In contrast, we follow seminal work on the paradigmatic metallic magnet Cr [33, 61] in which even weak interactions trigger a transition to an incommensurate antiferromagnetic state due to a special nesting-property of the band structure. Given the electronic structure of ferropnictides, it is natural to assume that antiferromagnetic order emerges, at least partly, due to near nesting between the dispersions of holes and electrons. Assuming weak to moderate interactions we advocate an itinerant spin density wave scenario for iron based superconductors [1, 9, 14, 36, 37, 56, 73].

On the other hand, we show how to calculate the spin excitations in the magnetically ordered state for the itinerant magnetic. In contrast to localized spin models this is considerably more involved task as we have to deal with spin and charge degrees of freedom simultaneously. Hence, we first review the formalism for the single-band Hubbard model in the simple (π, pi) antiferromagnetic state. Then we move to the situation of multi-band ferropnictides with stripe-like magnetism. The presence of additional orbital degrees of freedom make the theory more involved but in turn give rise to new interesting phenomena. The competing nesting vectors at $(\pi, 0)$ and $(0, \pi)$ provide a source of a new kind of *frustration* which is normally only studied in insulating magnets. Similar to the case of frustrated spin-only models our *itinerant* description finds accidental zero modes related to the competing ground states. Note that we do not discuss the details of the competition between antiferromagnetism and superconductivity in iron-based superconductors as this subject was discussed in detail during previous schools [50].

2.2 A Primer: Spin Density Wave and Spin Waves in the Single-Band Hubbard Model

As a forerunner of the itinerant weak-coupling description of antiferromagnetism in the multi-band iron-based systems we start with the simpler example of a single-band Hubbard model at half-filling which was analyzed in the context of parent high-T_c cuprates [8, 63, 66]:

$$H_1 = \sum_{\mathbf{k},\sigma} \varepsilon_{\mathbf{k}} c^{\dagger}_{\mathbf{k},\sigma} c_{\mathbf{k},\sigma} + \frac{U}{V} \sum_{\mathbf{k},\mathbf{k}',\mathbf{q}} c^{\dagger}_{\mathbf{k}+\mathbf{q},\uparrow} c_{\mathbf{k},\uparrow} c^{\dagger}_{\mathbf{k}'-\mathbf{q},\downarrow} c_{\mathbf{k}',\downarrow}. \tag{2.1}$$

Here, $c^{\dagger}_{\mathbf{k},\sigma}(c_{\mathbf{k},\sigma})$ is the creation (annihilation) of a fermionic quasiparticle with momentum $\mathbf{k}$ and spin σ. U denotes the on-site Coulomb repulsion and $\varepsilon_{\mathbf{k}}$ is the dispersion based on the tight-binding approximation with hopping between nearest neighbors. At half filling ($n = 1$) the Fermi surface is perfectly nested, i.e. $\varepsilon_{\mathbf{k}} = -\varepsilon_{\mathbf{k}+\mathbf{Q}}$ where $\mathbf{Q} = (\pi, \pi)$. Similar to the one-dimensional case the non-interacting susceptibility $\chi_0(\mathbf{Q}, \omega)$ diverges at the nesting wave vector. It signals that the paramagnetic ground state is unstable towards the formation of density waves. For repulsive Coulomb interaction the spin density wave instability occurs first and the system undergoes the paramagnetic $|PM\rangle$ to antiferromagnetic $|AF\rangle$ transition for any weak value of U. In a mean-field approximation we decouple the interaction with respect to a finite magnetization. Putting the spin quantization axis along the z-direction we define the spin density wave gap

$$\Delta = \frac{U}{V} \sum_{\mathbf{k},\sigma}{}' \operatorname{sgn}(\sigma) \langle AF | c^{\dagger}_{\mathbf{k}+\mathbf{Q},\sigma} c_{\mathbf{k},\sigma} | AF \rangle. \tag{2.2}$$

The mean-field Hamiltonian is bilinear and can be diagonalized. This gives two new Eigenenergies $E_{1/2} = \pm\sqrt{\varepsilon_{\mathbf{k}}^2 + \Delta^2}$. They look similar to the ones of a superconductor with two bands separated by an energy gap 2Δ. Therefore, the system is an insulator for any small value of U at half-filling and complete nesting. If the nesting is not complete the spin density wave transition will occur only at some critical value of U_{cr}. Note, it is an artefact of the simple mean-field description that the system is only an insulator at exactly half-filling for any U.

Tpin excitations of the antiferromagnetic state are calculated from the dynamical spin susceptibility tensor

$$\chi^{st}(\mathbf{q}, \mathbf{q}', t) \propto \langle AF|T_\tau S^s_{\mathbf{q}}(t) S^t_{-\mathbf{q}'}(0)|AF\rangle \tag{2.3}$$

where $S^s_{\mathbf{q}} = \sum_{\mathbf{k},\alpha,\beta} c^\dagger_{\mathbf{k}+\mathbf{q},\alpha} \sigma^s_{\alpha,\beta} c_{\mathbf{k},\beta}$. It is important to calculate the fluctuations with respect to the true antiferromagnetic ground state with finite Δ. The spin rotational symmetry of the Hubbard Hamiltonian is explicitly broken in the antiferromagnetic state making the longitudinal (zz) and transverse ($+-$) components of the susceptibility tensor inequivalent. In particular, the zz component is gapped by 2Δ and is less interesting. The gapless Goldstone modes and corresponding spin waves appear in the transverse channel and govern the low energy behavior of the system. The transverse susceptibility $\chi^{\pm}(\mathbf{q}, \mathbf{q}', t)$ can be computed within a random phase approximation in the antiferromagnetic ground state. This amounts to summing all ladder diagrams in a Dyson equation:

$$\chi^{\pm}_{RPA}(\mathbf{q}, \mathbf{q}', \omega) = \sum_{\mathbf{q}_1} \chi^{\pm}_0(\mathbf{q}, \mathbf{q}_1, \omega)[1 - U\chi^{\pm}_0(\mathbf{q}_1, \mathbf{q}', \omega)]^{-1} \tag{2.4}$$

The translational symmetry is broken which reduces the size of the Brillouin zone by half. Hence, new off-diagonal components of the susceptibility appear in momentum space (Umklapp susceptibility). Their presence is crucial for the correct description of spin excitations. Moreover, in order to treat correctly gapless spin waves in the charge gap set by 2Δ, it is necessary to compute the magnitude of the gap, (2.2), and random phase approximation spin susceptibility self-consistently. Note, the condition for the pole formation in the random phase approximation transverse susceptibility for $\mathbf{q} = \mathbf{q}' = \mathbf{Q}$ and $\omega \to 0$ (Goldstone mode) coincides with the spin density wave mean-field gap equation [63].

These calculations of the spin waves – despite the apparent limitations because of the the assumption that fermionic excitations continue to be infinitely long lived quasiparticles even for large U – were carried out in application to the undoped cuprate high-T_c superconductors. Results of such a spin wave calculation are for example shown in Fig. 2.1 [58] together with experimental data [11]. The agreement is remarkable given the strongly correlated nature of cuprates. It is interesting to notice that the random phase approximation formalism taken in the large U limit gives the spin wave velocity $v = \frac{1}{\sqrt{2}}J$ with $J = -\frac{4t^2}{U}$ in agreement with strong-coupling result of the corresponding Heisenberg model. This indicates that at least at half-filling the spin excitations can be consistently described by both weak- and strong-coupling approaches. From the weak-coupling perspective, an inclusion of self-energy corrections [66] further improves the agreement between Monte-Carlo calculations of the sublattice magnetization and that obtained within the self-

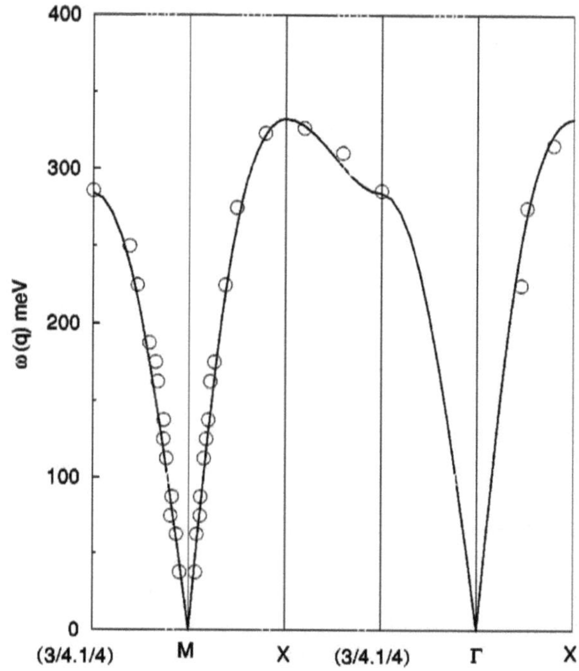

Fig. 2.1 Calculated spin wave dispersion using the self-consistent random phase approximation type formalism for the single band Hubbard model in comparison to experimental data for cuprates of Coldea et al. [11]. Taken from Peres and Araujo [58]

consistent random phase approximation type theory. It points to the idea that there is indeed a smooth crossover between weak and strong coupling limits [27], with the proper inclusion of higher order corrections.

The presented weak-coupling theory via a random phase approximation gives not only immediately information about spin excitations but also knowledge on the particle-hole continuum and its interaction with the spin waves. In the opposite strong-coupling limit this requires extra care. However, it is important to keep in mind that the weak-coupling theory has its own limitations because the underlying assumption that the transition between strong and weak-coupling is continuous is not true in general, in particular at small doping.

Having introduced some key ideas with a single-band system, let us now move to ferropnictides. The aim of the next sections is to generalize the single-band theory to the multiband, or multiorbital, case in order to gain insights into the magnetic phase of iron-based superconductors. An additional motivation is a desire to understand the magnetic properties of models which are most commonly used to describe the superconducting phase. This is required in order to obtain a coherent understanding of all different phases within a single approach. Moreover, in contrast to cuprates there is a principal difficulty to derive the analog of an effective low energy spin Hamiltonian for multiorbital systems such as ferropnictides in a controlled way.

2.3 Selection of Magnetic Order in the Parent Phase of Iron Based Superconductors

2.3.1 Magnetic Frustration

As pointed out in the introduction the magnetic order in the FeAs materials was originally detected by neutron scattering [13] and μsR experiments [35]. The magnetic transition temperature varies slightly from compound to compound and is of the order of $T_N \sim 150\,\text{K}$. In real space the magnetic ordering consists of ferromagnetic chains along one crystallographic direction in the Fe square lattice which are coupled antiferromagnetically. In momentum space the order can be characterized either by $\mathbf{Q}_1 = (\pi, 0)$ or $\mathbf{Q}_2 = (0, \pi)$ wavevector. Within the localized scenario this order emerges in the context of the $J_1 - J_2$ model [5] for $J_2 > 0.5J_1$ once quantum fluctuations are taken into account. In the following we review how this order appears in the itinerant picture using the simplified Fermi surface topology of the iron-based superconductors and the on-site interactions, written in the band space.

The schematic Fermi surface (FS) in the normal state of ferropnictides is depicted in Fig. 2.2 for the unfolded iron-based Brillouin zone. Here for simplicity we ignore the orbital flavors of the Fermi surface. As we pointed out in the introduction the elliptic electron bands and nearly circular hole bands are nearly nested.

For the Fermi surface topology of ferropnictides it means that there are two nesting wave vectors $\mathbf{Q}_1 = (\pi, 0)$ and $\mathbf{Q}_2 = (0, \pi)$ between the hole- and electron-like pockets. For the idealized case of zero ellipticity of the electron pockets and equal masses for the electron and hole bands the situation is similar to the half-filled Hubbard model with nearest neighbor hopping. In particular, the susceptibility in the particle-hole channel diverges logarithmically as it usually does in the

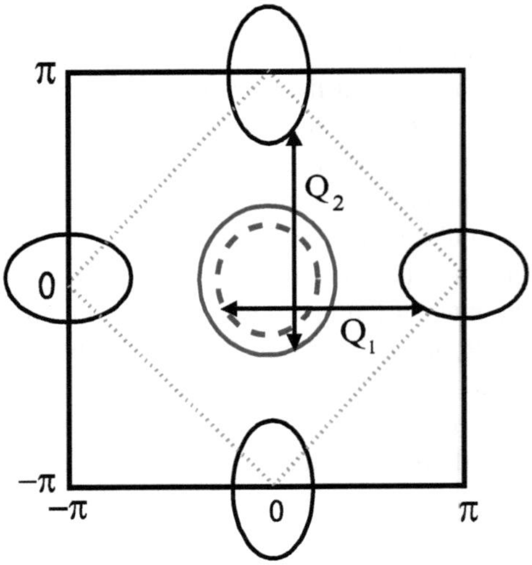

Fig. 2.2 Schematic Fermi surface of ferropnictides in the unfolded iron-based Brillouin zone with two circular hole-like pockets centered around the Γ-point and two elliptical electron-like pockets located around the $(\pi, 0)$ and $(0, \pi)$ point of the Brillouin zone, respectively. $\mathbf{Q}_1$ and $\mathbf{Q}_2$ represent two nesting wave vectors. Taken from Eremin and Chubukov [19]

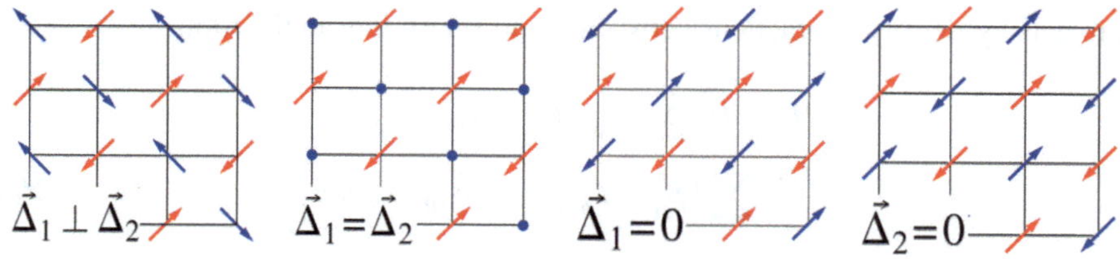

Fig. 2.3 Possible real space orderings of the magnetic state. Either one of the last two corresponds to the experimentally realized one. Taken from Eremin and Chubukov [19]

particle-particle Cooper-channel. A renormalization group analysis shows that the leading instability is magnetic. Chubukov et al. [9], Cvetkovic and Tesanovic [14] However, in contrast to the Hubbard model, here it occurs at two wave vectors $\mathbf{Q}_1$ and $\mathbf{Q}_2$ leaving the question open how the experimentally observed state is realized.

To formulate the problem in a formal way let us start with a generic spin configuration described by two mean-field spin density wave order parameters $\overrightarrow{\Delta}_i$ for each of the wave vector $\mathbf{Q}_i$ in the form

$$\overrightarrow{S}(\mathbf{R}) = \overrightarrow{\Delta}_1 e^{i\mathbf{Q}_1 R} + \overrightarrow{\Delta}_2 e^{i\mathbf{Q}_2 R}. \tag{2.5}$$

For such a configuration the Fe lattice decouples into two antiferromagnetically ordered sublattices with magnetizations $\overrightarrow{\Delta}_1 + \overrightarrow{\Delta}_2$ and $\overrightarrow{\Delta}_1 - \overrightarrow{\Delta}_2$, respectively, which interpenetrate each other. However, neither the angle between two Neel lattices is fixed nor the magnitude. For example, in Fig. 2.3 we show four possible orderings for a generic $\overrightarrow{S}(\mathbf{R})$ out of many possibilities. The last two configurations with one of the $\overrightarrow{\Delta}_i$ going to zero refer to the experimentally observed one.

Without loss of generality one can assume that one of the hole pockets is interacting stronger with the two electron-like pockets. Therefore it is useful to consider a model consisting of a single circular hole FS centered around the Γ-point (α-band) and two elliptical electron Fermi surface pockets centered around $(\pm\pi, 0)$ and $(0, \pm\pi)$ points in the unfolded Brillouin zone (β-bands):

$$H_2 = \sum_{\mathbf{p},\sigma} \left[\varepsilon_{\mathbf{p}}^{\alpha_1} \alpha_{1\mathbf{p}\sigma}^{\dagger} \alpha_{1\mathbf{p}\sigma} + \varepsilon_{\mathbf{p}}^{\beta_1} \beta_{1\mathbf{p}\sigma}^{\dagger} \beta_{1\mathbf{p}\sigma} + \varepsilon_{\mathbf{p}}^{\beta_2} \beta_{2\mathbf{p}\sigma}^{\dagger} \beta_{2\mathbf{p}\sigma} \right]. \tag{2.6}$$

Here, $\varepsilon_{\mathbf{p}}^{\alpha_1} = -\frac{\hbar^2 p^2}{2m_1} + \mu$ and $\varepsilon_{\mathbf{p}}^{\beta_1} = \frac{\hbar^2 p_x^2}{2m_x} + \frac{\hbar^2 p_y^2}{2m_y} - \mu$, $\varepsilon_{\mathbf{p}}^{\beta_2} = \frac{\hbar^2 p_x^2}{2m_y} + \frac{\hbar^2 p_y^2}{2m_x} - \mu$ are the dispersions of hole and electron bands. The momenta of $\alpha-$ fermions are counted from $(0, 0)$, the momenta of β_1- and β_2-fermions are counted from $(0, \pi)$ and $(\pi, 0)$.

One assumes that the most important interactions are those which give rise to the spin density wave order. According to the terminology of Chubukov et al. [9] these are density-density interactions between α and β fermions in the form

$$H_4 = U_1 \sum \alpha^{\dagger}_{1\mathbf{p}_3\sigma}\beta^{\dagger}_{j\mathbf{p}_4\sigma'}\beta_{j\mathbf{p}_2\sigma'}\alpha_{1\mathbf{p}_1\sigma} + \frac{U_3}{2}\sum\left[\beta^{\dagger}_{j\mathbf{p}_3\sigma}\beta^{\dagger}_{j\mathbf{p}_4\sigma'}\alpha_{1\mathbf{p}_2\sigma'}\alpha_{1\mathbf{p}_1\sigma} + h.c\right]. \tag{2.7}$$

At this level we neglect the potential angular dependencies of U_1 and U_3 along the FSs which arise due to hybridization of different orbitals which we will discuss later.

The two spin density wave order parameters are expressed as $\vec{\Delta}_1 \propto \sum_{\mathbf{p}} \langle\alpha^{\dagger}_{1\mathbf{p}\delta}\beta_{1\mathbf{p}\gamma}\sigma_{\delta\gamma}\rangle$ with momentum $\mathbf{Q}_1$ and $\vec{\Delta}_2 \propto \sum_{\mathbf{p}}\langle\alpha^{\dagger}_{1\mathbf{p}\delta}\beta_{2\mathbf{p}\gamma}\sigma_{\delta\gamma}\rangle$ with momentum $\mathbf{Q}_2$. Without loss of generality we can set $\vec{\Delta}_1$ along z-axis and $\vec{\Delta}_2$ in the xz-plane. In explicit form, we introduce

$$\Delta_1^z = -U_{SDW}\sum_{\mathbf{p}}\langle\alpha^{\dagger}_{1\mathbf{p}\uparrow}\beta_{1\mathbf{p}\uparrow}\rangle$$
$$\Delta_2^{z(x)} = -U_{SDW}\sum_{\mathbf{p}}\langle\alpha^{\dagger}_{1\mathbf{p}\uparrow}\beta_{2\mathbf{p}\uparrow(\downarrow)}\rangle. \tag{2.8}$$

where $U_{SDW} = U_1 + U_3$.

In the simplest case all masses are equal, i.e., $m_x = m_y$ and $\varepsilon^{\beta_1}_{\mathbf{p}} = \varepsilon^{\beta_2}_{\mathbf{p}} = \varepsilon^{\beta}_{\mathbf{p}}$, such that we have perfect nesting between all circular pockets. Again similar to the single-band case we can treat the interaction term, (2.7), with the spin density wave order parameters, (2.8) in mean-field. With two subsequent Bogolyubov transformations [19] the quadratic Hamiltonian can be written as

$$H_2^{\rm eff} = \sum_{a,\mathbf{p}} \varepsilon^{\beta}_{\mathbf{p}} d^{\dagger}_{a\mathbf{p}}d_{a\mathbf{p}} + \sum_{p} E_{\mathbf{p}}\left(e^{\dagger}_{a\mathbf{p}}e_{a\mathbf{p}} + p^{\dagger}_{b\mathbf{p}}p_{b\mathbf{p}} - e^{\dagger}_{b\mathbf{p}}e_{b\mathbf{p}} - p^{\dagger}_{a\mathbf{p}}p_{a\mathbf{p}}\right), \tag{2.9}$$

where $E_{\mathbf{p}} = \pm\sqrt{(\varepsilon_{\mathbf{p}})^2 + |\Delta|^2}$ and $\Delta = \sqrt{(\Delta_1^z)^2 + (\Delta_2)^2}$, $\Delta_2 = \sqrt{(\Delta_2^z)^2 + (\Delta_2^x)^2}$ and there is only one single self-consistent equation for the total gap magnitude

$$1 = \frac{U_{SDW}}{2N}\sum_{\mathbf{p}}\frac{1}{\sqrt{(\varepsilon_{\mathbf{p}})^2 + \Delta^2}}. \tag{2.10}$$

As one clearly sees the self-consistency (2.10) only fixes the total magnitude of $(\Delta_1^z)^2 + (\Delta_2)^2$ but not the magnitude and the direction of individual Δ_1 and Δ_2. This implies a huge groundstate degeneracy at the mean-field level where spin density wave ordering corresponds to spontaneous breaking of the $O(6)$ symmetry. The experimentally realized states with either $\Delta_i = 0$ are just two of infinitely many possibilities. Moreover, the degeneracy of the itinerant model is even larger than that in the localized $J_1 - J_2$ model where the moments are fixed. In the itinerant picture the magnitude of each sublattice magnetization can vary by the cost of the other.

In the following we will show that within mean-field the degeneracy can still be lifted by the ellipticity of the electron pockets or interaction within electron or within hole pockets. This explains why the particular order is realized and then due to magnetostriction the structural order is imposed on the lattice. However, it does not explain why the structural transition occurs sometimes at higher temperature than T_N. To cover this aspect we also briefly discuss how the wave vector can be selected prior to the onset of long-range magnetic ordering. This refers to the notion of Ising nematic order. We also discuss how it couples to the lattice and orbitals.

2.3.2 Lifting the Magnetic Ground State Degeneracy at T_N

So far, the analysis assumed the idealized situation of fully nested circular electron and hole pockets with interactions only between holes and electrons. If one takes into account the ellipticity of the electron pockets and additional interactions the degeneracy may be lifted. One includes, in addition, the four other possible $\beta - \beta$ interactions:

$$\begin{aligned}
H_4^{ex} = & \qquad\qquad\qquad\qquad\qquad\qquad\qquad\qquad (2.11)\\
& U_6 \sum \beta^\dagger_{1\mathbf{p}_3\sigma}\beta^\dagger_{2\mathbf{p}_4\sigma'}\beta_{2\mathbf{p}_2\sigma'}\beta_{1\mathbf{p}_1\sigma} + U_7 \sum \beta^\dagger_{2\mathbf{p}_3\sigma}\beta^\dagger_{1\mathbf{p}_4\sigma'}\beta_{2\mathbf{p}_2\sigma'}\beta_{1\mathbf{p}_1\sigma} \\
& + \frac{U_8}{2} \sum \left[\beta^\dagger_{2\mathbf{p}_3\sigma}\beta^\dagger_{2\mathbf{p}_4\sigma'}\beta_{1\mathbf{p}_2\sigma'}\beta_{1\mathbf{p}_1\sigma} + h.c\right] + \\
& \frac{U_4}{2} \sum \left[\beta^\dagger_{1\mathbf{p}_3\sigma}\beta^\dagger_{1\mathbf{p}_4\sigma'}\beta_{1\mathbf{p}_2\sigma'}\beta_{1\mathbf{p}_1\sigma} + \beta^\dagger_{2\mathbf{p}_3\sigma}\beta^\dagger_{2\mathbf{p}_4\sigma'}\beta_{2\mathbf{p}_2\sigma'}\beta_{2\mathbf{p}_1\sigma}\right]
\end{aligned}$$

In the antiferromagnetic state one again applies the sequence of Bogolyubov transformations, takes averages $\langle \cdots \rangle$, to obtain the contribution to the ground state energy. The final correction to the ground state energy was obtained in Eremin and Chubukov [19] and has the form:

$$\begin{aligned}
E_{gr}^{ex} = & 2A^2 \left[(U_6 + U_8 - U_7 - U_4)\right] \frac{|\Delta_1|^2|\Delta_2|^2}{\Delta^4} + \\
& 4A^2 U_7 \frac{(\Delta_1 \cdot \Delta_2)^2}{\Delta^4} \qquad (2.12)
\end{aligned}$$

Observe that E_{gr}^{ex} depends on $|\Delta_1|^2|\Delta_2|^2$ and on $(\Delta_1 \cdot \Delta_2)^2$, i.e., it is sensitive to both, relative values and relative directions of Δ_1 and Δ_2. When all interactions are of equal strength, the first term vanishes, and the last term favors $\Delta_1 \perp \Delta_2$. In this situation, the $O(6)$ degeneracy of the ideal model is broken, but only down to $O(3) \times O(2)$, i.e., the magnitude of the order parameter at each site is now the same because

$\left(\vec{\Delta}_1 + \vec{\Delta}_2\right)^2 = \left(\vec{\Delta}_1 - \vec{\Delta}_2\right)^2$, but the angle between the directions of the spin density wave order in the two sublattices (i.e., between $\vec{\Delta}_1 + \vec{\Delta}_2$ and $\vec{\Delta}_1 - \vec{\Delta}_2$) is still arbitrary. This is exactly the same situation as in the classical $J_1 - J_2$ model. However, once $U_6 + U_8 - U_7 - U_4$ is nonzero, the degeneracy is broken down to a conventional $O(3)$ already at the mean-field level. Because U_4 is reduced and even changes sign under RG [9], while other U_i do not flow, the most likely situation is that $U_6 + U_8 - U_7 - U_4 > 0$, in which case E_{gr}^{ex} is minimized when either $\vec{\Delta}_1 = 0$, or $\vec{\Delta}_2 = 0$, i.e., spin density wave order is either $(0, \pi)$ or $(\pi, 0)$. This is exactly the same spin density wave order as observed in experiments. If $U_6 + U_8 - U_7 - U_4$ was negative, E_{gr}^{ex} would be minimized when $|\vec{\Delta}_1| = |\vec{\Delta}_2|$, in which case the spin density wave order parameters of the two sublattices would align orthogonal to each other. The spin configuration for such state is shown the left panel of Fig. 2.3.

The other effect concerns a distortion of the electron Fermi pockets, i.e. the electron pockets are actually ellipses rather than circles. The effective masses m_x and m_y are not equal, and $\varepsilon_{\mathbf{k}}^{\beta_1} \neq \varepsilon_{\mathbf{k}}^{\beta_2}$. To continue with the analytical analysis, one assumes that the ellipticity is small, and introduces $m_x = (1+\delta)m$ and $m_y = (1-\delta)m$, where $\delta << 1$, and computes the correction to the ground state energy to second order in δ. The contribution to the ground state energy coming from the ellipticity is

$$E_{gr}^{ellipt} = C|\vec{\Delta}_1|^2|\vec{\Delta}_2|^2, \quad C = \delta^2 \frac{m\mu^2}{4\pi\Delta^4} \tag{2.13}$$

where the coefficient C is positive, i.e., the correction due to ellipticity of electron pockets again breaks the degeneracy and selects either $(0, \pi)$ or $(\pi, 0)$ state. This is again the same selection as one needs for consistency with experiments. It is remarkable that ellipticity leads to a term in the ground state energy that is similar to an effective interaction between two spin density wave order parameters which, for $\Delta << \mu$, leads to the same selection of the ground state spin density wave order as the direct interaction between the two electron pockets.

It follows from the above analysis that it is indeed possible to understand in a simple and transparent way the connection between the peculiar FS topology and the magnetic groundstate in ferropnictides. Before we continue to calculate the most striking manifestation of symmetry breaking in the antiferromagnetic state, the Goldstone modes, we shortly discuss different aspects. Since the magnetic $(\pi, 0)$ state not only breaks spin rotational symmetry but also translational symmetry it leads to a reconstruction of the Fermi surface when going from the paramagnetic to the antiferromagnetic phase. In Fig. 2.4 we show the data of this reconstruction (upper panel) from Liu et al. [45] together with a theoretical modeling of an itinerant five orbital model (lower panel). The agreement is remarkably good.

Note that a reconstruction of the electronic structure in the antiferromagnetic state leads to a corresponding modification of the local density of states. The latter shows a pronounced C_2 anisotropy in the antiferromagnetic state which is consistent with STM experiments. Chuang et al. [7] Fig. 2.5 shows a calculated real space image of

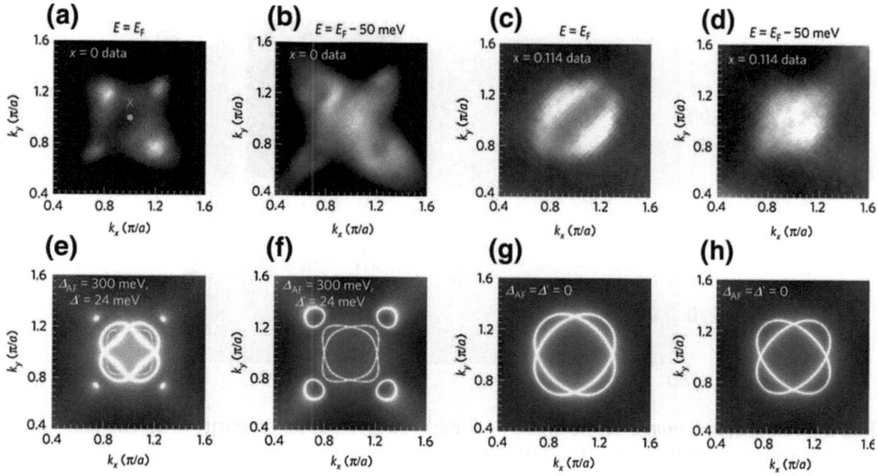

Fig. 2.4 Reconstruction of the Fermi surface pockets from the paramagnetic state, panels **c**, **d** and **g**, **h**, to the antiferromagnetic state, panels **a**, **b** and **e**, **f**. The angle-resolved photoemission data (*upper panels*) are nicely explained by the theoretical calculation of an itinernat five orbital model (*lower panel*). Taken from Liu et al. [45]

the quasiparticle intereference pattern (QPI) for a non-magnetic impurity computed within a T-matrix approach [37] away from the impurity position and at $-7\,\text{meV}$. Observe that the QPI modulation shown in the inset is reflected in periodic structures seen along the x and y crystallographic directions. In the antiferromagnetic x direction with $\mathbf{q}_1$ structure, one finds a weak stripe pattern with a periodicity of the order of $7a$ where a is a lattice (Fe-Fe) constant. In addition, there is a stronger modulation of $10 \div 15a$ along the FM y direction, consistent with the structure described as nematic in Chuang et al. [7]. These extra modulations arise purely due to scattering between bands involved in the spin density wave, no extra folding of the bands occurs.

In addition, two remarks are necessary. First, once the linear coupling term between lattice distortion and magnetization which breaks C(4) symmetry is included in the theory, it forces the lattice to follow the magnetic ordering [3], and thus, leads to a structural transition. Experimentally such a structural transition at T_s is observed but depending on the material it either coincides or precedes the magnetic transition at T_N.

The emergence of a phase with broken rotational symmetry above or at T_N is experimentally not only manifested by a tetragonal to orthorhombic transition at T_s, but also by the onset of significant anisotropies in several quantities [26], such as dc resistivity [6, 68], optical conductivity [18, 55], and orbital occupancy [78].

Early theoretical proposals explored two alternative scenarios for the nematic order. On the one hand orbital ordering might induce the structural transition and trigger the magnetic transition at a lower temperature by renormalizing the exchange constants. Kruger et al. [40] On the other hand, a purely magnetic scenario based on a localized $J_1 - J_2$ model [5], for ferropnictides see Si and Abrahams [64], Fang

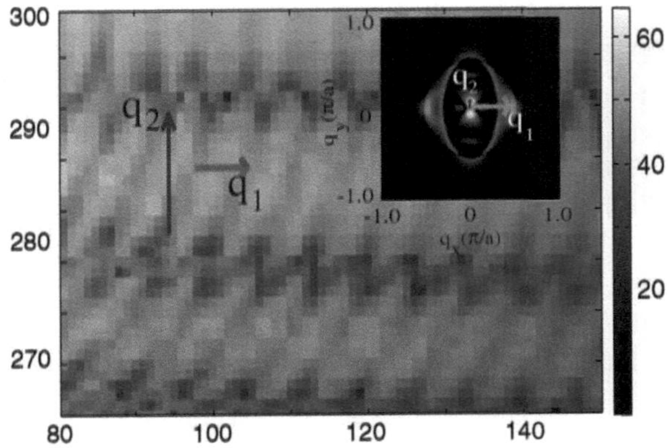

Fig. 2.5 Real space image of the QPI at −7 meV away from the impurity site. The inset shows the corresponding QPI image. The QPI interference originating from the structure in the spectral density is reflected in the strong $10 \div 15a$ modulation along the FM (y) direction and a weak $5 \div 7a$ modulation along the antiferromagnetic (x) direction. The impurity potential is a mixture of magnetic (20%) and nonmagnetic (80%) parts. Intensity refers to states/eV. Taken from Knolle et al. [37]

et al. [21], Xu et al. [75], is build upon the assumption that the order parameter manifold for two possible stripe magnetic orders is O(3)$\times Z_2$ (compare to the actual degeneracy which has O(6) symmetry), with the Z_2 Ising symmetry differentiating between the two possible ordering vectors $(0, \pi)$ and $(\pi, 0)$. This scenario explores the possibility that the Z_2 Ising variable orders before the O(3) rotational symmetry is broken, leading to an intermediate phase with a broken tetragonal symmetry but no long-range magnetic order, called Ising nematic. This magnetic scenario was later extended to itinerant systems in a phenomenological way [22, 57]. Subsequently, we have shown that such an Ising nematic order can also be understood within a magnetic scenario for an itinerant fermionic model. Fernandes et al. [23] The formalism is beyond the scope of these notes but we recommend the interested reader to consult the reviews on this subject [24].

2.4 Orbital Structure of Spin Density Wave State

In case of the iron-based superconductors, the electrons of the $3d$-states of the t_{2g} manifold, which are d_{xz}, d_{yz} and d_{xy} orbitals, determine the spectral density of the electronic states at low energies and introduce orbital differentiation on the Fermi surface pockets. In the remainder of the section, we will study its consequences for the spin-density wave state and the different superconducting gap symmetries.

The kinetic part of the Hamiltonian is defined in the orbital basis of the xz, yz and xy orbitals, which contains intra and interorbital hoppings. Upon diaginalization of the kinetic part of the Hamiltonian this underlying orbital structure leads to a momentum structure of the interaction written on the band level. In order to get a basic feeling for the orbital structure and especially to see the effect of the orbital degrees of freedom we concentrate on the simplest multioribital model that was introduce in the context of ferropnictides [59]. It takes into account only the d_{xz} and d_{yz} orbitals which, however, have the largest weight around the Fermi energy.

The kinetic part of the Hamiltonian is written in the two component spinor basis in orbital space with the help of the Pauli matrices τ_i

$$\mathscr{H}_0 = \sum_{\mathbf{k},\sigma} \psi_\sigma^\dagger \left[(\varepsilon_+ - \mu)\, \tau_0 + \varepsilon_- \tau_3 + \varepsilon_{xy} \tau_1 \right] \psi_\sigma \tag{2.14}$$

where

$$\psi_\sigma(\mathbf{k}) = \begin{pmatrix} d_{xz\sigma}(\mathbf{k}) \\ d_{yz\sigma}(\mathbf{k}) \end{pmatrix}. \tag{2.15}$$

Here, $\varepsilon_\pm(\mathbf{k}) = \frac{\varepsilon_x(\mathbf{k}) \pm \varepsilon_y(\mathbf{k})}{2}$, $\varepsilon_x(\mathbf{k}) = -2t_1 \cos k_x - 2t_2 \cos k_y - 4t_3 \cos k_x \cos k_y$, $\varepsilon_y(\mathbf{k}) = -2t_2 \cos k_x - 2t_1 \cos k_y - 4t_3 \cos k_x \cos k_y$, and $\varepsilon_{xy}(\mathbf{k}) = -4t_4 \sin k_x \sin k_y$.

In general, the interacting part of the Hamiltonian consists of different on-site electron-electron interactions,

$$\begin{aligned} H_{int}^{orb} = {} & U \sum_i \sum_\nu n_{i\nu\uparrow} n_{i\nu\downarrow} + V \sum_{\nu \neq \mu, \sigma, \sigma'} n_{i\nu\sigma} n_{i\mu\sigma'} - \\ & J \sum_{\nu \neq \mu} \mathbf{S}_{i\nu} \cdot \mathbf{S}_{i\mu} + J' \sum_{\nu \neq \mu} d_{i\nu\uparrow}^\dagger d_{i\nu\downarrow}^\dagger d_{i\mu\downarrow} d_{i\mu\uparrow}, \end{aligned} \tag{2.16}$$

where U and V refer to the intra- and inter-orbital Coulomb repulsion, J and $J' = J/2$ denote the Hund and the pair hopping terms. The definition of the spin operators is

$$\mathbf{S}_{i\nu} = \frac{1}{2} \sum_{\sigma\sigma'} d_{i\nu\sigma}^\dagger \sigma_{\sigma\sigma'} d_{i\nu\sigma'}, \qquad \sigma = (\sigma_1, \sigma_2, \sigma_3) \text{ Pauli matrices.} \tag{2.17}$$

The first term accounts for Coulomb repulsion between two electrons which have the same quantum numbers in orbital description up to the electron spin. These electrons are located at the same lattice site and belong to the same Fe $3d$-orbital so that the overlap of both wave functions in real space is large. Because of this, intra-orbital Coulomb repulsion U is the strongest interaction of H_{int}. The second term denoted with U' has a similar meaning but the electrons belong to different orbitals. As a consequence, the inter-orbital Coulomb repulsion U' is usually weaker than U as the orbital overlap is smaller. The third term is inter-orbital exchange J for electrons at the same site but from different orbitals ν and μ. The interpretation of the last term with coupling J' is simultaneous hopping of two electron from one orbital to another orbital at the same site. These are on-site interactions that are easily generalized for a multi orbital model.

It is easy to see that, in the band basis, fermions from different bands can have the same or a different orbital character, so that the strength of the repulsion varies between U, V and so on. Because the orbital content of the bands is determined by the diagonalization of the kinetic part of the Hamiltonian, so are the interactions of fermions in the band basis that arise from the underlying Coulomb repulsion in the orbital basis. This introduces a momentum dependence of the band interactions and it will be the first task to rewrite the interactions in terms of band operator. Furthermore, a projection on the low-energy states, i.e. the Fermi surface pockets, will be required.

2.4.1 Transformation from Orbital to Band Basis

So far, the spin density wave state has been calculated within in the band basis assuming some effective intra- and inter-band interactions. At this stage we perform the unitary transformations to see the impact of the orbitals on the spin-density wave order. The similarity transformation is

$$\begin{pmatrix} d_{xz\mathbf{k}\sigma} \\ d_{yz\mathbf{k}\sigma} \end{pmatrix} = \begin{pmatrix} \cos\vartheta_{\mathbf{k}\sigma} & \sin\vartheta_{\mathbf{k}\sigma} \\ -\sin\vartheta_{\mathbf{k}\sigma} & \cos\vartheta_{\mathbf{k}\sigma} \end{pmatrix} \begin{pmatrix} \alpha_{\mathbf{k}\sigma} \\ \beta_{\mathbf{k}\sigma} \end{pmatrix} \tag{2.18}$$

where the operators $\alpha^\dagger_{\mathbf{k}\sigma}$ and $\beta^\dagger_{\mathbf{k}\sigma}$ create fermions in the hole-like and electron-like band. The tight-binding Hamiltonian $\mathscr{H}_0$ (2.14) is diagonal in the new basis

$$\mathscr{H}_0 = \sum_{\mathbf{k}\sigma} E_{\alpha\mathbf{k}\sigma} n_{\alpha\mathbf{k}\sigma} + E_{\beta\mathbf{k}\sigma} n_{\beta\mathbf{k}\sigma} \tag{2.19}$$

Before the canonical transformation (2.18) is applied, the quartic terms I-IV are Fourier transformed from the real space to momentum space, e.g., the intra-orbital Coulomb repulsion is readily rewritten

$$\sum_{i\nu\sigma} d^\dagger_{i\nu\sigma} d^\dagger_{i\nu\bar\sigma} d_{i\nu\bar\sigma} d_{i\nu\sigma} = \sum_{\substack{\mathbf{k}+\mathbf{k}'=\mathbf{k}''+\mathbf{k}''' \\ \nu\sigma}} d^\dagger_{\nu\mathbf{k}\sigma} d^\dagger_{\nu\mathbf{k}'\bar\sigma} d_{\nu\mathbf{k}''\bar\sigma} d_{\nu\mathbf{k}'''\sigma} \tag{2.20}$$

where conservation of momentum is implied. In a next step, this becomes in band space

$$\begin{aligned} &\sum_{\nu\sigma} d^\dagger_{\nu\mathbf{k}\sigma} d^\dagger_{\nu\mathbf{k}'\bar\sigma} d_{\nu\mathbf{k}''\bar\sigma} d_{\nu\mathbf{k}'''\sigma} = \\ &\sum_{\sigma} (c_{\mathbf{k}}\alpha^\dagger_{\mathbf{k}\sigma} + s_{\mathbf{k}}\beta^\dagger_{\mathbf{k}\sigma})(c_{\mathbf{k}'}\alpha^\dagger_{\mathbf{k}'\bar\sigma} + s_{\mathbf{k}'}\beta^\dagger_{\mathbf{k}'\bar\sigma})(c_{\mathbf{k}''}\alpha_{\mathbf{k}''\bar\sigma} + s_{\mathbf{k}''}\beta_{\mathbf{k}''\bar\sigma})(c_{\mathbf{k}'''}\alpha_{\mathbf{k}'''\sigma} + s_{\mathbf{k}'''}\beta_{\mathbf{k}'''\sigma}) \\ &\quad +(s_{\mathbf{k}}\alpha^\dagger_{\mathbf{k}\sigma} - c_{\mathbf{k}}\beta^\dagger_{\mathbf{k}\sigma})(s_{\mathbf{k}'}\alpha^\dagger_{\mathbf{k}'\bar\sigma} - c_{\mathbf{k}'}\beta^\dagger_{\mathbf{k}'\bar\sigma})(s_{\mathbf{k}''}\alpha_{\mathbf{k}''\bar\sigma} - c_{\mathbf{k}''}\beta_{\mathbf{k}''\bar\sigma})(s_{\mathbf{k}'''}\alpha_{\mathbf{k}'''\sigma} - c_{\mathbf{k}'''}\beta_{\mathbf{k}'''\sigma}) \end{aligned} \tag{2.21}$$

where $c_{\mathbf{k}} = \cos\vartheta_{\mathbf{k}\sigma}$ and $s_{\mathbf{k}} = \sin\vartheta_{\mathbf{k}\sigma}$. The term on the right hand side of (2.21) comprises all different electron-hole interactions in band space which corresponds to intraorbital Coulomb repulsion in orbital space.

However, only some of these terms will drive the system towards a Fermi surface instability. For example in case of a a spin density wave state at commensurate ordering momenta $\mathbf{Q} = X$ and $\mathbf{Q} = Y$, which roughly connects hole and electron pockets, the averages

$$\langle \alpha^{\dagger}_{\mathbf{k}\sigma}\beta_{\mathbf{k}+\mathbf{Q}\sigma}\rangle \quad \text{and} \quad \langle \alpha^{\dagger}_{\mathbf{k}\sigma}\beta_{\mathbf{k}+\mathbf{Q}\bar{\sigma}}\rangle \tag{2.22}$$

are finite. The expectation value on the left hand side is the analog of magnetization out-of-plane and the other one is the analog to magnetization in-plane. One task will thus be project the interactions to the relevant band channels which contribute to a magnetism and superconductivity (Fig. 2.6).

The channels contributing to spin density wave and superconducting order in the band basis are given by [9]

$$\begin{aligned}
u_1^{\sigma\sigma'}: &\quad \alpha^{\dagger}_{\mathbf{k}\sigma}\beta^{\dagger}_{\mathbf{k}'\sigma'}\beta_{\mathbf{k}''\sigma'}\alpha_{\mathbf{k}'''\sigma} \\
u_2^{\sigma\sigma'}: &\quad \alpha^{\dagger}_{\mathbf{k}\sigma}\beta^{\dagger}_{\mathbf{k}'\sigma'}\alpha_{\mathbf{k}''\sigma'}\beta_{\mathbf{k}'''\sigma} \\
u_3^{\sigma\sigma'}: &\quad \alpha^{\dagger}_{\mathbf{k}\sigma}\alpha^{\dagger}_{\mathbf{k}'\sigma'}\beta_{\mathbf{k}''\sigma'}\beta_{\mathbf{k}'''\sigma} + \text{h.c.} \\
u_4^{\sigma\sigma'}: &\quad \alpha^{\dagger}_{\mathbf{k}\sigma}\alpha^{\dagger}_{\mathbf{k}'\sigma'}\alpha'_{\mathbf{k}''\sigma'}\alpha'_{\mathbf{k}'''\sigma} + \text{h.c.} \\
u_5^{\sigma\sigma'}: &\quad \beta^{\dagger}_{\mathbf{k}\sigma}\beta^{\dagger}_{\mathbf{k}'\sigma'}\beta'_{\mathbf{k}''\sigma'}\beta'_{\mathbf{k}'''\sigma} + \text{h.c.}
\end{aligned} \tag{2.23}$$

where the hermitian conjugate of interaction channel $u_{3,4,5}$ are added explicitly, because the second pair of operators is formally allowed to belong to a different Fermi surface pocket. Note that all types of Coulomb interactions (2.16) in general contribute to all five band channels, as is demonstrated (2.21).

In the following, we will discuss spin-density wave order and, without loss of generality, the antiferromagnetic wave-vector is set to $\mathbf{Q} = X$, which connects parts of the Fermi surface that belong to different bands. Assuming conservation of momentum $\mathbf{k} + \mathbf{k}' = \mathbf{k}'' + \mathbf{k}'''$ and remembering that hole pockets and electron pockets belong to hole band E_{α} and electron band E_{β} respectively, a first reduction of the number of different interactions can be achieved.

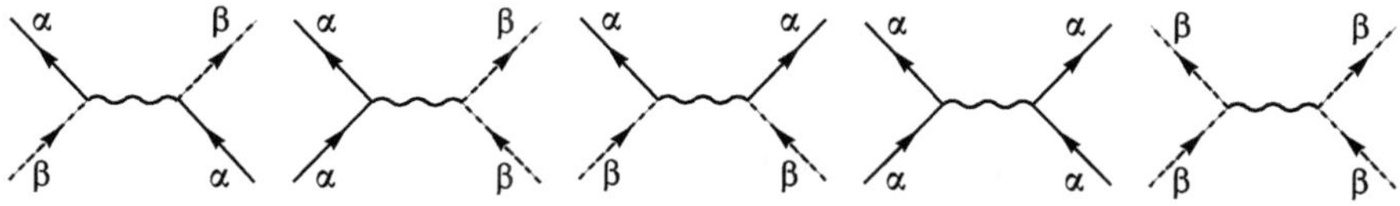

Fig. 2.6 From *left to right* Band channels u_1, u_2, u_3, u_4 and u_5. The operator $\alpha^{\dagger}$ ($\beta^{\dagger}$) creates a fermion in a hole-like (electron-like) band. Note that the first three channels give rise to a magnetic Fermi surface instability while the last three band channels promote superconductivity

$$
\begin{aligned}
u_1^{\sigma\sigma'}: &\quad \alpha^\dagger_{\mathbf{k}\sigma}\beta^\dagger_{\mathbf{k}'+\mathbf{Q}\sigma'}\beta_{\mathbf{k}+\mathbf{Q}\sigma'}\alpha_{\mathbf{k}'\sigma} \\
u_2^{\sigma\sigma'}: &\quad \alpha^\dagger_{\mathbf{k}\sigma}\beta^\dagger_{\mathbf{k}'+\mathbf{Q}\sigma'}\alpha_{\mathbf{k}'\sigma'}\beta_{\mathbf{k}+\mathbf{Q}\sigma} \\
u_3^{\sigma\sigma'}: &\quad \alpha^\dagger_{\mathbf{k}\sigma}\alpha^\dagger_{\mathbf{k}'\sigma'}\beta_{\mathbf{k}+\mathbf{Q}\sigma'}\beta_{\mathbf{k}'+\mathbf{Q}\sigma} + \text{h.c.} \\
&\quad \alpha^\dagger_{\mathbf{k}\sigma}\alpha^\dagger_{\mathbf{k}'\sigma'}\beta_{\mathbf{k}'+\mathbf{Q}\sigma'}\beta_{\mathbf{k}+\mathbf{Q}\sigma} + \text{h.c.}
\end{aligned}
\tag{2.24}
$$

Both quartics terms of channel u_3 seem identical up to a factor (-1) if $\sigma = \sigma'$ and thus seem to compensate each other. There is a caveat however, because the factor in front of the quartic terms due to band-orbital matrix elements is only equal in the case of the Coulomb interactions U and J'. In general, the factor has to be written down and evaluated for every term separately.

Looking closer at (2.24), channels that contribute to the mean field average $\langle\alpha^\dagger_{\mathbf{k}\sigma}\beta_{\mathbf{k}+\mathbf{Q}\sigma}\rangle$ and $\langle\alpha^\dagger_{\mathbf{k}\sigma}\beta_{\mathbf{k}+\mathbf{Q}\bar{\sigma}}\rangle$ can be identified

$$
\begin{aligned}
\text{magnetization out-of-plane } \langle\alpha^\dagger_{\mathbf{k}\sigma}\beta_{\mathbf{k}+\mathbf{Q}\sigma}\rangle: &\quad u_1^{\sigma\sigma}, u_2^{\sigma\sigma}, u_2^{\sigma\bar{\sigma}}, u_3^{\sigma\sigma}, u_3^{\sigma\bar{\sigma}} \\
\text{magnetization in-plane } \langle\alpha^\dagger_{\mathbf{k}\sigma}\beta_{\mathbf{k}+\mathbf{Q}\bar{\sigma}}\rangle: &\quad u_1^{\sigma\bar{\sigma}}, u_3^{\sigma\bar{\sigma}}
\end{aligned}
\tag{2.25}
$$

2.4.2 Spin Density Wave Mean Field Equations

In this section, the system of coupled mean field equation will be derived. Therefore, the Hamiltonian H_{int} is projected on the interactions channels $u_1^{\sigma\sigma'}$, $u_2^{\sigma\sigma'}$ and $u_3^{\sigma\sigma'}$. This leads to a mean field Hamiltonian with two parts that relate to spin density wave order with magnetization out-of-plane and in-plane. Both parts are obtained in the same way. The Hamiltonian $\mathscr{H} = \mathscr{H}_0 + \mathscr{H}_{\text{SDW}}$ is given by

$$
\mathscr{H}_0 = \sum_{\mathbf{k}\sigma} E_{\alpha\mathbf{k}\sigma} n_{\alpha\mathbf{k}\sigma} + E_{\beta\mathbf{k}\sigma} n_{\beta\mathbf{k}\sigma} \tag{2.26}
$$

and

$$
\begin{aligned}
\mathscr{H}_{\text{SDW}} = &\sum_{\mathbf{k},\mathbf{k}'\sigma} \Gamma^{\sigma\bar{\sigma}}_{\text{SDW}}(\mathbf{k},\mathbf{k}')\langle\alpha^\dagger_{\mathbf{k}'\sigma}\beta_{\mathbf{k}'+\mathbf{Q}\bar{\sigma}}\rangle\alpha^\dagger_{\mathbf{k}\sigma}\beta_{\mathbf{k}+\mathbf{Q}\bar{\sigma}} \\
&+ \sum_{\mathbf{k},\mathbf{k}'\sigma} \Gamma^{\sigma\sigma}_{\text{SDW}}(\mathbf{k},\mathbf{k}')\langle\alpha^\dagger_{\mathbf{k}'\sigma}\beta_{\mathbf{k}'+\mathbf{Q}\sigma}\rangle\alpha^\dagger_{\mathbf{k}\sigma}\beta_{\mathbf{k}+\mathbf{Q}\sigma} \\
&+ \text{h.c.}
\end{aligned}
\tag{2.27}
$$

where the order parameter are defined as

$$
\Delta_{\mathbf{k}}^{\sigma\sigma'} = \sum_{\mathbf{k}} \Gamma^{\sigma\sigma'}_{\text{SDW}}(\mathbf{k},\mathbf{k}')\langle\alpha^\dagger_{\mathbf{k}'\sigma}\beta_{\mathbf{k}'+\mathbf{Q}\sigma'}\rangle \tag{2.28}
$$

The potentials $\Gamma^{\sigma\sigma'}_{\mathrm{SDW}}(k, k')$ should be the same because both magnetizations are equivalent as was already pointed out. Furthermore, the momentum dependence is a consequence of the band-orbital matrix elements. In case where one magnetization is set to zero, the Hamiltonian H can be written in the form

$$\mathscr{H} = \sum_{\mathbf{k}\sigma} \left(\alpha^{\dagger}_{\mathbf{k}\sigma}, \beta^{\dagger}_{\mathbf{k}+\mathbf{Q}\sigma'}\right) \begin{pmatrix} E_{\alpha\mathbf{k}\sigma} & \Delta^{\sigma\sigma'}_{\mathbf{k}} \\ \Delta^{\sigma\sigma'}_{\mathbf{k}} & E_{\beta\mathbf{k}+\mathbf{Q}\sigma'} \end{pmatrix} \begin{pmatrix} \alpha_{\mathbf{k}\sigma} \\ \beta_{\mathbf{k}+\mathbf{Q}\sigma'} \end{pmatrix} \tag{2.29}$$

with either $\sigma' = \sigma$ (out-of-plane) or $\sigma' = \bar{\sigma}$ (in-plane). It can be diagonalized with a standard Bogoliubov transformation. The renormalized energy bands are

$$E_{(\pm)\mathbf{k}\sigma} = \frac{E_{\alpha\mathbf{k}\sigma} + E_{\beta\mathbf{k}+\mathbf{Q}\sigma'}}{2} \pm \sqrt{\left(\frac{E_{\alpha\mathbf{k}\sigma} - E_{\beta\mathbf{k}+\mathbf{Q}\sigma'}}{2}\right)^2 + \left(\Delta^{\sigma\sigma'}_{k}\right)^2} \tag{2.30}$$

and the order parameter is easily evaluated in the new basis and satisfies the self-consistent equation

$$\Delta^{\sigma\sigma'}_{\mathbf{k}} = \frac{1}{2}\sum_{\mathbf{k}'} \Gamma^{\sigma\sigma'}_{\mathrm{SDW}}(\mathbf{k}, \mathbf{k}') \frac{\Delta^{\sigma\sigma'}_{\mathbf{k}'}}{E_{(+)\mathbf{k}'} - E_{(-)\mathbf{k}'}} \left(\tanh\left(\frac{E_{(+)\mathbf{k}'}}{2k_B T}\right) - \tanh\left(\frac{E_{(-)\mathbf{k}'}}{2k_B T}\right)\right) \tag{2.31}$$

So far, it was only assumed that the potentials inside the sum are real and spin independent functions which are invariant under $k \leftrightarrow k'$. Exploiting the fact that the potentials are determined by the transformation from orbital to band space, one finds

$$\begin{aligned} \Gamma^{\sigma\sigma}_{\mathrm{SDW}}(\mathbf{k}, \mathbf{k}') &= 2\Big[-u^{\sigma\sigma}_1(\mathbf{k}, \mathbf{k}') + u^{\sigma\sigma}_2(\mathbf{k}, \mathbf{k}') - u^{\sigma\bar{\sigma}}_2(\mathbf{k}, \mathbf{k}') - u^{\sigma\sigma}_3(\mathbf{k}, \mathbf{k}') + u^{\sigma\bar{\sigma}}_3(\mathbf{k}, \mathbf{k}')\Big] \\ \Gamma^{\sigma\bar{\sigma}}_{\mathrm{SDW}}(\mathbf{k}, \mathbf{k}') &= 2\Big[-u^{\sigma\bar{\sigma}}_1(\mathbf{k}, \mathbf{k}') - u^{\sigma\bar{\sigma}}_3(\mathbf{k}, \mathbf{k}')\Big] \end{aligned} \tag{2.32}$$

where $\langle\alpha^{\dagger}_{\mathbf{k}\sigma}\beta_{\mathbf{k}+\mathbf{Q}\sigma}\rangle = -\langle\alpha^{\dagger}_{\mathbf{k}\bar{\sigma}}\beta_{\mathbf{k}+\mathbf{Q}\bar{\sigma}}\rangle$ and $\langle\alpha^{\dagger}_{\mathbf{k}\sigma}\beta_{\mathbf{k}+\mathbf{Q}\bar{\sigma}}\rangle = \langle\alpha^{\dagger}_{\mathbf{k}\sigma}\beta_{\mathbf{k}+\mathbf{Q}\bar{\sigma}}\rangle$ is already taken into account. Every such potentials contains terms of every Coulomb interactions (2.16). They are given explicitly in Table 2.1 together with Table 2.2.

Finally, the spin-density wave potentials are

$$\Gamma^{\sigma\sigma'}_{\mathrm{SDW}}(\mathbf{k}, \mathbf{k}') = -2Ua_1 + \left(U' + \tfrac{J}{2} + J'\right)(a_4 - a_3) + 2Ja_2 \tag{2.33}$$

where the momentum dependence is carried by the coefficients a_i which are defined in Table 2.2. As has been expected, both potentials $\Gamma^{\sigma\sigma}_{\mathrm{SDW}}$ and $\Gamma^{\sigma\bar{\sigma}}_{\mathrm{SDW}}$ for magnetization out-of-plane and in-plane respectively turn out to be identical.

Though, one task remains until the potentials can be used to solve the self-consistent equation of the spin density wave order parameter (2.32). Obviously, an order parameter that satisfies $\Delta_{\mathbf{k}} = -\Delta_{-\mathbf{k}}$ breaks time-reversal symmetry and is not allowed in general. Consequently, the potential and thus the coefficients a_i, which consist of products of band-orbital matrix elements of the transformation (2.34), need to be invariant under the operation $\mathbf{k} \rightarrow -\mathbf{k}$ and $\mathbf{k}' \rightarrow -\mathbf{k}'$.

Table 2.1 Interactions $u^{\sigma\sigma'}_{1,2,3}$. The first four potentials on the left hand side contribute to magnetization out-of-plane and the other two potentials contribute to in-plane magnetization

$u_1^{\sigma\sigma}(\mathbf{k},\mathbf{k}') = +\frac{U}{2}a_1$	$+\left(\frac{U'}{2}+\frac{J}{4}\right)a_3$	$-\frac{J}{2}a_2$	$-\frac{J'}{2}a_4$
$u_2^{\sigma\sigma}(\mathbf{k},\mathbf{k}') = +\frac{U}{2}a_1$	$-\left(\frac{U'}{2}+\frac{J}{4}\right)a_2$	$+\frac{J}{2}a_3$	$-\frac{J'}{2}a_4$
$u_2^{\sigma\bar{\sigma}}(\mathbf{k},\mathbf{k}') = +\frac{U}{2}a_1$	$-\left(\frac{U'}{2}+\frac{J}{4}\right)a_2$	$+\frac{J}{2}a_3$	$-\frac{J'}{2}a_4$
$u_3^{\sigma\sigma}(\mathbf{k},\mathbf{k}') = +\frac{U}{2}(a_1-a_1)$	$+\left(\frac{U'}{2}+\frac{J}{4}\right)(a_2-a_4)$	$+\frac{J}{2}(a_4-a_2)$	$+\frac{J'}{2}(a_3-a_3)$
$u_3^{\sigma\bar{\sigma}}(\mathbf{k},\mathbf{k}') = -\frac{U}{2}a_1$	$+\left(\frac{U'}{2}+\frac{J}{4}\right)a_2$	$+\frac{J}{2}a_4$	$-\frac{J'}{2}a_3$
$u_1^{\sigma\bar{\sigma}}(\mathbf{k},\mathbf{k}') = +\frac{U}{2}a_1$	$+\left(\frac{U'}{2}+\frac{J}{4}\right)a_3$	$-\frac{J}{2}a_2$	$-\frac{J'}{2}a_4$
$u_3^{\sigma\bar{\sigma}}(\mathbf{k},\mathbf{k}') = +\frac{U}{2}a_1$	$-\left(\frac{U'}{2}+\frac{J}{4}\right)a_4$	$-\frac{J}{2}a_2$	$+\frac{J'}{2}a_3$

Table 2.2 Products of band-orbital matrix elements occurring due to the transformation from orbital to band space

$a_1 = s_{\mathbf{k}}c_{\mathbf{k}+\mathbf{Q}}s_{\mathbf{k}'}c_{\mathbf{k}'+\mathbf{Q}} + c_{\mathbf{k}}s_{\mathbf{k}+\mathbf{Q}}c_{\mathbf{k}'}s_{\mathbf{k}'+\mathbf{Q}}$
$a_2 = s_{\mathbf{k}}s_{\mathbf{k}+\mathbf{Q}}s_{\mathbf{k}'}s_{\mathbf{k}'+\mathbf{Q}} + c_{\mathbf{k}}c_{\mathbf{k}+\mathbf{Q}}c_{\mathbf{k}'}c_{\mathbf{k}'+\mathbf{Q}}$
$a_3 = s_{\mathbf{k}}c_{\mathbf{k}+\mathbf{Q}}c_{\mathbf{k}'}s_{\mathbf{k}'+\mathbf{Q}} + c_{\mathbf{k}}s_{\mathbf{k}+\mathbf{Q}}s_{\mathbf{k}'}c_{\mathbf{k}'+\mathbf{Q}}$
$a_4 = s_{\mathbf{k}}s_{\mathbf{k}+\mathbf{Q}}c_{\mathbf{k}'}c_{\mathbf{k}'+\mathbf{Q}} + c_{\mathbf{k}}c_{\mathbf{k}+\mathbf{Q}}s_{\mathbf{k}'}s_{\mathbf{k}'+\mathbf{Q}}$

This is assured by defining

$$\begin{pmatrix}\alpha_{\mathbf{k}\sigma}\\ \beta_{\mathbf{k}\sigma}\end{pmatrix} \equiv \operatorname{sgn}\cos\vartheta_{\mathbf{k}\sigma}\begin{pmatrix}\cos\vartheta_{\mathbf{k}\sigma} & -\sin\vartheta_{\mathbf{k}\sigma}\\ \sin\vartheta_{\mathbf{k}\sigma} & \cos\vartheta_{\mathbf{k}\sigma}\end{pmatrix}\begin{pmatrix}d_{xz\mathbf{k}\sigma}\\ d_{yz\mathbf{k}\sigma}\end{pmatrix} \tag{2.34}$$

for ordering momentum $\mathbf{Q} = X$ and for $\mathbf{Q} = Y$,

$$\begin{pmatrix}\alpha_{\mathbf{k}\sigma}\\ \beta_{\mathbf{k}\sigma}\end{pmatrix} \equiv \operatorname{sgn}\sin\vartheta_{\mathbf{k}\sigma}\begin{pmatrix}\cos\vartheta_{\mathbf{k}\sigma} & -\sin\vartheta_{\mathbf{k}\sigma}\\ \sin\vartheta_{\mathbf{k}\sigma} & \cos\vartheta_{\mathbf{k}\sigma}\end{pmatrix}\begin{pmatrix}d_{xz\mathbf{k}\sigma}\\ d_{yz\mathbf{k}\sigma}\end{pmatrix} \tag{2.35}$$

which is always allowed at every momentum k as the rows of the matrix of the unitary transformation are only determined up to a complex phase.

With this definition, it is possible to get a first estimate of the resulting spin density wave potential Γ_{SDW}. We know that the character of a fermion on electron pocket e_2 centered around Y is (xz). Hence, the band-orbital matrix elements near e_2 are $s_{k+Q} \approx 1$ and $c_{k+Q} \approx 0$ whereas at hole pocket h_1 the matrix elements are $s_k \approx |\sin\varphi|$ and $s_k \approx \operatorname{sgn}\sin\varphi\cos\varphi$. With this approximation, the potential (2.33) becomes for momenta k and k' at hole pocket h_1 and electron pocket e_2

$$\Gamma^{\sigma\sigma'}_{\mathrm{SDW}}(\varphi_{\mathbf{k}},\varphi_{\mathbf{k}'}) = -\operatorname{sgn}(\sin\varphi\sin\varphi')(2U\cos\varphi\cos\varphi' - J\sin\varphi\sin\varphi') \tag{2.36}$$

Remind that the interaction potential does not depend on interorbital Coulomb repulsion and pair hopping term as it is the case in orbital description. Usually, the Coulomb interaction U will be an order of magnitude larger than the Hund term J and the potential with momenta along the hole pocket is approximately

$$|\Gamma_{\mathrm{SDW}}^{\sigma\sigma'}(\varphi_{\mathbf{k}}, \varphi_{\mathbf{k}'})| \propto |\cos\varphi\cos\varphi'| \tag{2.37}$$

Likewise, one can show that the potential along hole pocket h_2 will be described by the same expression (2.37). This is reasonable because both hole pockets are centered around Γ in the real Brilloun zone and thus should have a similar gap.

Since the sum of the right hand side of the self-consistent gap (2.28) is governed by terms with $E_{\alpha k\sigma} - E_{\beta k+Q\sigma'} \approx 0$ the shape of the gap will be determined by the potential (2.37). Thus, a natural ansatz for the gap function in the two orbital model when taking into account the band-orbital matrix elements is

$$|\Delta_{\varphi_{\mathbf{k}}}^{\sigma\sigma'}| = |\cos\varphi_{\mathbf{k}}| \tag{2.38}$$

where $\varphi_{\mathbf{k}}$ is the angle measured along a hole pocket relative to the k_x-axis. Apparently, this order parameter has nodes due to underlying orbital structure, which is rather different from the expectation of having the simple isotropic magnetic gap used in the previous section.

2.4.3 Numerical Results

In orbital description, nodes have been found close to the Fermi level of the renormalized bands for a constant order parameter. It is thus eligible to expect nodes in the band description, too. In band description, however, only if the gap is very small the Fermi surface topology will be reproduced because of poor nesting [38]. In addition, no nodal points of the band structure appear. Solving the mean field equation with the anisotropic gap, which arises naturally due to the transformation from orbital to band space as was outlined in the previous section, the Fermi surface pockets are not completely gaped (Fig. 2.7). In other words, the simple reason for the remnant pockets at the Fermi surface is that the order parameter is nodal. Furthermore, the shape of the gap does not depend on the parameter V and J', because they give rise to a sub-leading solution of the mean-field equations which is orthogonal to the leading one, i.e., corresponds to different harmonics. The order parameter resembles a $|\cos\varphi|$ near the Fermi level and the contour lines of the Fermi surface143 pockets.

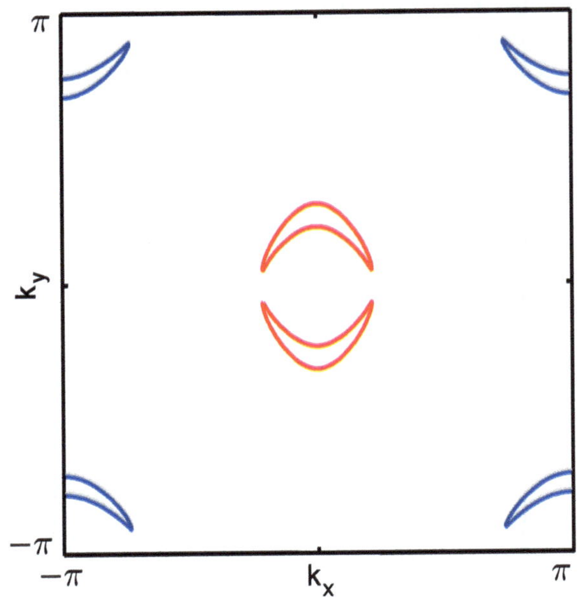

Fig. 2.7 Fermi surface of the spin-density wave state at half-filling derived within the band description for $U = 4.0|t|, J = 0.2U$ *Red* electron-like pockets. *Blue* hole-like pockets

Here, the angle φ correspond to momentum $(k_x, k_y) = (k_F \cos\varphi, k_F \sin\varphi)$ where φ is measured from the k_x-axis in positive direction.

The nodal structure of the renormalized bands is shown for $U = 4.0|t|$ and $U = 5.0|t|$ which leads to a gap which is comparable with the gap in orbital description. It was demonstrated that the angle-dependent nodal gap is a direct consequence of the orbital character of the two Fe $3d$-orbitals. The sign changing character of the orbital lobes leads to an inter-band exchange energy in tight-binding representation which inevitably has nodes. The band-orbital matrix elements of the similarity transformation from orbital to band space are then in approximately $\sin\varphi_{\mathbf{k}}$ and $\cos\varphi_{\mathbf{k}}$. Consequently, the rewritten Hubbard-type interactions in band space have nodes and so does the gap. Furthermore, not just the Fermi surface topology was obtained in orbital and band description but the nodal structure of the bands close to the Fermi level, too. This illustrates that the correct way to describe the spin density wave state in band description is to take into account the angle-dependence of the interaction potentials due to the underlying orbital character of the conduction electrons.

2.5 Orbital Effects for the Superconducting State

We recast the minimal Hamiltonian that is sufficient to describe a superconducting system, which will be treated within a mean-field approach,

$$\mathcal{H}_{\mathrm{SC}} = \sum_{\mathbf{k}\sigma} \xi(\mathbf{k}) \alpha^{\dagger}_{\mathbf{k}\sigma} \alpha_{\mathbf{k}\sigma} + \sum_{\mathbf{k}\sigma} \Gamma_{\mathrm{SC}}(\mathbf{k}, \mathbf{k}') \alpha^{\dagger}_{\mathbf{k}\sigma} \alpha^{\dagger}_{-\mathbf{k}\bar{\sigma}} \alpha_{-\mathbf{k}'\bar{\sigma}} \alpha_{\mathbf{k}'\sigma}, \tag{2.39}$$

and which includes only singlet Cooper-pairs. The interaction $\Gamma_{\mathrm{SC}}(\mathbf{k}, \mathbf{k}')$ is often approximated as momentum and frequency independent and the dispersion $\xi(\mathbf{k})$ describes low-energy states near the Fermi level of a single-band model. In particular, the states which fulfil the relation $\xi(\mathbf{k}) = 0$ define the corresponding Fermi surface pocket of this energy band. Consequently, the minimal Hamiltonian is just applicable in systems where one energy band crosses the Fermi level, for example Cuprate superconductors. In the following, we will extend the Hamiltonian such that it is able to describe a system with momentum dependent interactions and several Fermi surface pockets. Such a superconductor is a multi-condensate superconductor. The most prominent are of course the ferropnictides, which are not just high-temperature superconductors, but which also have rich fermiology. Moreover, we will introduce the leading angular harmonics approximation [47] to deal with the momentum dependence of the interactions and the set of coupled gap equations and point out the importance for several types of superconductors with different non-trivial gap symmetries, such as extended s-wave, d-wave and p-wave superconductors. For completeness, we also give the self-consistent gap equation of the problem (2.39), after the Hamiltonian was reduced to an effective one-particle Hamiltonian within mean-field,

$$\Delta(\mathbf{k}) = -\sum_{\mathbf{k}'} \Gamma_{\mathrm{SC}}(\mathbf{k}, \mathbf{k}') \langle \alpha^{\dagger}_{\mathbf{k}'\uparrow} \alpha^{\dagger}_{-\mathbf{k}'\downarrow} \rangle, \qquad \Delta(\mathbf{k}) \in \mathbb{R}. \tag{2.40}$$

In case of inter-pocket interaction, the momentum transfer is large so that a scattering of a Cooper-pair from one pocket to another is possible as shown in Fig. 2.8.

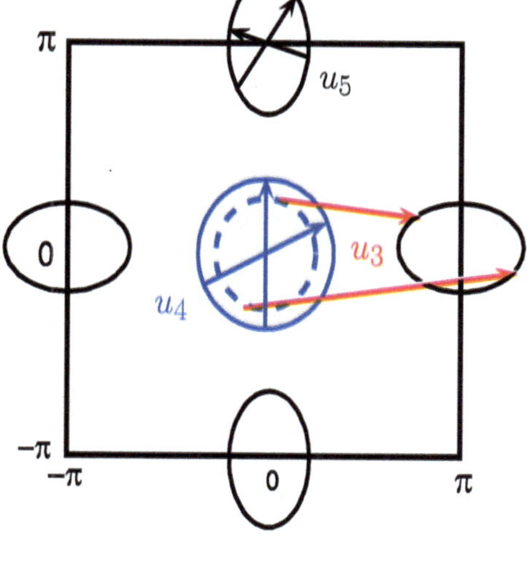

Fig. 2.8 Band channels that contribute to superconductivity. While channel u_4 and u_5 refer to intra-pocket scattering for hole and electron Fermi surface pockets, channel u_3 corresponds to inter-pocket scattering

In the band basis, we use operators $\alpha^\dagger_{\mathbf{k}\sigma}$ and $\beta^\dagger_{\mathbf{k}+\mathbf{Q}\sigma}$ for fermions that belong to a hole and electron Fermi surface pocket, respectively. Here, the shift in momentum by $\mathbf{Q}$ can corresponds to either $\mathbf{Q} = \mathrm{X}$ or $\mathbf{Q} = \mathrm{Y}$ as usual. We again list the band channels explicitly,

$$\begin{aligned} u_3: &\quad \alpha^\dagger_{\mathbf{k}\sigma}\alpha^\dagger_{\mathbf{k}'\sigma'}\beta_{\mathbf{k}''\sigma'}\beta_{\mathbf{k}'''\sigma} \;+\; \text{h.c.},\\ u_4: &\quad \alpha^\dagger_{\mathbf{k}\sigma}\alpha^\dagger_{\mathbf{k}'\sigma'}\alpha_{\mathbf{k}''\sigma'}\alpha_{\mathbf{k}'''\sigma},\\ u_5: &\quad \beta^\dagger_{\mathbf{k}\sigma}\beta^\dagger_{\mathbf{k}'\sigma'}\beta_{\mathbf{k}''\sigma'}\beta_{\mathbf{k}'''\sigma}. \end{aligned} \tag{2.41}$$

In the following, we restrict superconductivity to singlet- and intra-band pairing, such that especially the momentum is fixed at the Fermi level to the contour lines of the Fermi surface pockets as sketched in Fig. 2.8. In particular, this implies $\mathbf{k} = -\mathbf{k}'$ and $\mathbf{k}'' = -\mathbf{k}'''$ for (2.41). Therefore, we are left with the task to determine the interactions potentials that correspond to the mean-field contributions of band channels u_3, u_4 and u_5, which we will denote $\Gamma_{\mathrm{he}}(\mathbf{k},\mathbf{k}')$, $\Gamma_{\mathrm{hh}}(\mathbf{k},\mathbf{k}')$ and $\Gamma_{\mathrm{ee}}(\mathbf{k},\mathbf{k}')$ accordingly. Here, we used the index 'h' to mark a Cooper-pair sitting on a hole-like band and 'e' for a Cooper-pair sitting on an electron-like band. The interaction potentials correspond to Hubbard U, V, J and J' times the orbital-dressing that accounts for the orbital character of the fermions of hole and electron Fermi surface pockets. Namely, when the orbital operators $d^\dagger$ of the Coulomb interaction are expressed by a linear combination of band operators $\alpha^\dagger$ and $\beta^\dagger$, one only needs to gather the correct terms that contribute to the respective band channels in order to obtain the corresponding so-called orbital-dressing factors. For every quartic term, there will be four of these band-orbital matrix elements. Moreover, because there are four distinct Hubbard interactions, we will need to repeat the process for all of them and add up the results. The one-particle Hamiltonian will then have the following form in the multi-band case

$$\begin{aligned} \mathscr{H}_{\mathrm{SC}} &= \sum_{i,\mathbf{k}\sigma} \xi_{\mathrm{h}_i}(\mathbf{k})\alpha^\dagger_{i\mathbf{k}\sigma}\alpha_{i\mathbf{k}\sigma} + \sum_{j,\mathbf{k}\sigma} \xi_{\mathrm{e}_j}(\mathbf{k})\beta^\dagger_{j\mathbf{k}\sigma}\beta_{j\mathbf{k}\sigma} \\ &+ \left(\sum_{i,\mathbf{k}\sigma} \Delta_{\mathrm{h}_i}(\mathbf{k})\alpha_{i-\mathbf{k}\sigma}\alpha_{i\mathbf{k}\bar{\sigma}} + \sum_{j,\mathbf{k}\sigma} \Delta_{\mathrm{e}_j}(\mathbf{k})\beta_{j-\mathbf{k}\sigma}\beta_{j\mathbf{k}\bar{\sigma}} + \text{h.c.} \right). \end{aligned} \tag{2.42}$$

where the pendant of the gap equation for the single band case (2.40) is,

$$\begin{aligned} \Delta_{\mathrm{h}_i}(\mathbf{k}) &= -\sum_{i'\mathbf{k}'} \Gamma_{\mathrm{h}_i\mathrm{h}_{i'}}(\mathbf{k},\mathbf{k}')\langle \alpha^\dagger_{i'\mathbf{k}'\uparrow}\alpha^\dagger_{i'-\mathbf{k}'\downarrow}\rangle - \sum_{i'\mathbf{k}'} \Gamma_{\mathrm{h}_i\mathrm{e}_{j'}}(\mathbf{k},\mathbf{k}')\langle \beta^\dagger_{j'\mathbf{k}'\uparrow}\beta^\dagger_{j'-\mathbf{k}'\downarrow}\rangle,\\ \Delta_{\mathrm{e}_j}(\mathbf{k}) &= -\sum_{i'\mathbf{k}'} \Gamma_{\mathrm{e}_j\mathrm{h}_{i'}}(\mathbf{k},\mathbf{k}')\langle \alpha^\dagger_{i'\mathbf{k}'\uparrow}\alpha^\dagger_{i'-\mathbf{k}'\downarrow}\rangle - \sum_{j'\mathbf{k}'} \Gamma_{\mathrm{e}_j\mathrm{e}_{j'}}(\mathbf{k},\mathbf{k}')\langle \beta^\dagger_{j'\mathbf{k}'\uparrow}\beta^\dagger_{j'-\mathbf{k}'\downarrow}\rangle. \end{aligned} \tag{2.43}$$

Fortunately, the superconducting Hamiltonian $\mathscr{H}_{\mathrm{SC}}$ can still be solved by separate Bogoliubov transformations, however, and most importantly, the superconducting gaps or condensates are coupled at the level of the self-consistent equations. The reason are the non-vanishing terms that mix Fermi surface pockets with different index with each other. In the limit where only intra-pocket scattering is possible, i.e. only $\Gamma_{\mathrm{h}_i\mathrm{h}_{i'}}(\mathbf{k},\mathbf{k}') \neq 0$ and $\Gamma_{\mathrm{e}_j\mathrm{e}_{j'}}(\mathbf{k},\mathbf{k}') \neq 0$ and only if $i = i'$ and $j = j'$, one deals with a situation where the system possesses several independent superconducting condensates. The critical superconducting temperature will then be dictated by the condensate with highest value of T_c. We note however, that the Coulomb repulsion is repulsive, so that the interaction potentials will also be positive quantities, such that the self-consistent equations of the form (2.40) only have the trivial solution, which is no superconductivity at all down to $T = 0$. In fact, we will demonstrate how the negative sign on the right-hand side of the self-consistent gap equations can be overcome in the case where the different condensates are coupled, even if the interactions remains repulsive. This is only possible if at least two gaps have a non-vanishing relative phase. Especially in the case where inter-pocket scattering (band channel u_3) is dominating, this means that the negative sign on the right hand side of the self-consistent gap equations is compensated by a sign change (or relative phase of 180°) between the different gaps, which can be a so-called $s^{\pm}$-state. Similarly, the repulsive interactions can be overcome by strong oscillation of the gap function or d-wave superconductors. We will consider various scenarios below with the help of leading angular harmonics approximation to treat the multi-condensate superconductivity on different Fermi surface pockets.

In order to solve the self-consistent gap equations, we note that superconductivity in the band basis is a Fermi surface instability. Therefore, we need to focus on the low-energy states and we restrict the summation over states to those directly at the Fermi level. This has the following advantages. For the Fermi surface pockets, the position of the Cooper-pairs sufficiently defined by angle $\varphi = \arctan k_y/k_x$ and $\varphi' = \arctan k_y'/k_x'$ and we can use $(k_x, k_y) \to (k, \varphi)$. Moreover, in a 2D system, the density-of-states and thus the contribution of each Fermi surface pocket to the superconducting condensate does not depend on the size of the pockets. It only depends on the effective mass of the corresponding energy dispersion, which is parabolic,

$$\xi_{h_i} = \mu_{h_i} - \frac{k^2}{2m_{h_i}}, \qquad \xi_{e_j} = \frac{k^2}{2m_{e_j}} - \mu_{e_j}. \tag{2.44}$$

If we restrict the self-consistent equation to the Fermi level and study the onset of the superconducting gap, the self-consistent equation becomes linear in the gap function,

$$\begin{aligned}\lambda\Delta_{\mathrm{h}_i}(\varphi) = &-\sum_{i'}\int_0^{2\pi}\frac{d\varphi'}{2\pi}\Gamma_{\mathrm{h}_i\mathrm{h}_{i'}}(\varphi,\varphi')\Delta_{h_{i'}}(\varphi')N_{h_{i'}}L_{h_{i'}}\\ &-\sum_{j}\int_0^{2\pi}\frac{d\varphi'}{2\pi}\Gamma_{\mathrm{h}_i\mathrm{e}_j}(\varphi,\varphi')\Delta_{e_j}(\varphi')N_{e_j}L_{e_j},\end{aligned} \tag{2.45}$$

$$\lambda\Delta_{e_j}(\varphi) = -\sum_i \int_0^{2\pi} \frac{d\varphi'}{2\pi} \Gamma_{e_j h_i}(\varphi, \varphi')\Delta_{h_i}(\varphi')N_{h_i}L_{h_i}$$
$$-\sum_{j'} \int_0^{2\pi} \frac{d\varphi'}{2\pi} \Gamma_{e_j e_{j'}}(\varphi, \varphi')\Delta_{e_{j'}}(\varphi')N_{e_{j'}}L_{e_{j'}}, \tag{2.46}$$

so that we need to introduce the eigenvalue λ. The indexes i and j correspond to a Fermi surface pocket each. The self-consistent gap equations for the electron Fermi surface pockets are the same. We introduced the density of states at the Fermi level, which explicitly is proportional to the effective mass $N_{h_{i'}} = m_{h_{i'}}/2\pi$, and the result of the energy integration $L_{h_{i'}} = \ln E_{F,h_{i'}}/T_c$. We also implied $(k_F, \varphi) \to (\varphi)$ everywhere. In case where the effective mass and the Fermi energy E_F do not vary much between the different Fermi surface pockets, we can pull it in front of the integral and can approximately use $N_{h_i}L_{h_i} \approx N_{e_j}L_{e_j}$. Instead, we can also remove these factors completely and absorb them in the interactions potentials.

2.5.1 Leading Angular Harmonics Approximation

Next, the gap function and the interactions potentials are treated within leading angular harmonics approximation and the coupled set of self-consistent gap equations will become a feasible eigenvalue problem. Hence, it is necessary to write down the basis functions of the irreducible representations for the tetragonal lattice symmetry [10, 51]. The relevant irreducible representations are A_{1g} (s-wave), B_{1g} ($d_{x^2-y^2}$-wave) and B_{2g} (d_{xy}-wave). The corresponding leading basis function, i.e. leading harmonics, are given in Table 2.3. At the Fermi level, we can approximate the Fermi surface pockets as circular sheets for the moment. We also assume that k_F is small. Near Γ and near X, where we first shift by $k_x \to k_x + \mathbf{Q}$, the expansion gives the following table. Note that Y is just related by a rotation of 90°. In this step, we summed up the contributions from expansion of the different harmonics. We stress that the coefficients in front of the harmonics are not determined by the expansion in k_F, but are independent and can be large in general. They are determined by the angular dependence of the interaction potentials that stem from the alternating orbital content. Since the different representations do not mix, the interactions potentials take the following form

$$\Gamma_{\mathrm{hh}}(\varphi, \varphi') = \sum_{m,n\geq 0} A_{\mathrm{hh}}^{mn} \cos 4m\varphi \cos 4n\varphi' + B_{\mathrm{hh}}^{mn} \cos((4m+2)\varphi) \cos((4n+2)\varphi')$$
$$+ C_{\mathrm{hh}}^{mn} \sin((4m+2)\varphi) \sin((4n+2)\varphi'), \tag{2.47}$$

Table 2.3 Basis functions of the corresponding superconducting gap symmetries. The angle is measured around different high-symmetry points, which each have a Fermi surface pocket centered around them. The angle $\varphi = 0$ points in the positive k_x-direction of the Brillouin zone. The coefficients in front of the harmonics are the degrees of freedom that are allowed to vary. The possible harmonics however are limited by the gap symmetry

s-wave	$d_{x^2-y^2}$-wave	d_{xy}-wave
$1, \cos k_x + \cos k_y, \cos k_x \cos k_y$	$\cos k_x - \cos k_y, \cos 2k_x - \cos 2k_y$	$\sin k_x \sin k_y$
$\Gamma\ \sum_{m\geq 0} a_m \cos 4m\varphi$	$\sum_{m\geq 0} b_m \cos((4m+2)\varphi)$	$\sum_{m\geq 0} c_m \sin((4m+2)\varphi)$
$X\ \sum_{n\geq 0} d_n \cos 2m\varphi$	$\sum_{n\geq 0} e_n \cos 2n\varphi$	$\sum_{m\geq 0} f_n \sin((4n+2)\varphi)$
$Y\ \sum_{n\geq 0} (-1)^n d_n \cos 2n\varphi$	$\sum_{n\geq 0} (-1)^n e_n \cos 2n\varphi$	$\sum_{n\geq 0} f_n \sin((4n+2)\varphi)$

$$\Gamma_{\mathrm{he}_1}(\varphi, \varphi') = \sum_{m,n\geq 0} \left\{A^{mn}_{\mathrm{he}} \cos 4m\varphi + B^{mn}_{\mathrm{he}} \cos((4m+2)\varphi)\right\} \cos 2n\varphi' + C^{mn}_{\mathrm{he}} \sin((4m+2)\varphi) \sin((4n+2)\varphi'), \tag{2.48}$$

$$\Gamma_{\mathrm{he}_2}(\varphi, \varphi') = \sum_{m,n\geq 0} (-1)^n \left\{A^{mn}_{\mathrm{he}} \cos 4m\varphi - B^{mn}_{\mathrm{he}} \cos((4m+2)\varphi)\right\} \cos 2n\varphi' + C^{mn}_{\mathrm{he}} \sin((4m+2)\varphi) \sin((4n+2)\varphi'). \tag{2.49}$$

Here, the hole Fermi surface pocket h is representative for every other pocket centered around Γ, while pocket e_1 is exclusively centered around X, while e_2 is centered around Y. The extension to more than those pockets h, e_1 and e_2 is straightforward. We also report the general form of the interaction potentials for intra-pocket scattering $\Gamma_{e_j e_j}$ and inter-pocket scattering $\Gamma_{e_j e_{j'}}$, where $j \neq j'$.

$$\Gamma_{\mathrm{e}_1\mathrm{e}_1}(\varphi, \varphi') = \sum_{m,n\geq 0} (A^{mn}_{\mathrm{ee}} + B^{mn}_{\mathrm{ee}}) \cos 2m\varphi \cos 2n\varphi' + C^{mn}_{\mathrm{ee}} \sin((4n+2)\varphi) \sin((4m+2)\varphi'), \tag{2.50}$$

$$\Gamma_{\mathrm{e}_2\mathrm{e}_2}(\varphi, \varphi') = \sum_{m,n\geq 0} (-1)^{m+n} (A^{mn}_{\mathrm{ee}} + B^{mn}_{\mathrm{ee}}) \cos 2m\varphi \cos 2n\varphi' + C^{mn}_{\mathrm{ee}} \sin((4n+2)\varphi) \sin((4m+2)\varphi'), \tag{2.51}$$

and

$$\Gamma_{e_1e_2}(\varphi,\varphi') = \sum_{m,n\geq 0} (-1)^n (A_{ee}^{mn} - B_{ee}^{mn}) \cos 2m\varphi \cos 2n\varphi' \\ + C_{ee}^{mn} \sin((4n+2)\varphi) \sin((4m+2)\varphi'), \quad (2.52)$$

$$\Gamma_{e_2e_1}(\varphi,\varphi') = \sum_{m,n\geq 0} (-1)^m (A_{ee}^{mn} - B_{ee}^{mn}) \cos 2m\varphi \cos 2n\varphi' \\ + C_{ee}^{mn} \sin((4n+2)\varphi) \sin((4m+2)\varphi'). \quad (2.53)$$

The form of the interactions potentials is still very general. They are merely restricted to s-wave and d-wave, which are the best candidates for almost all families of high-temperature superconductors. In the following, the superconducting gap for every Fermi surface pocket will be approximated by the leading angular harmonic. Again, the expansion of the interaction potentials above is valid for a system with hole Fermi surface pockets centered around Γ and electron Fermi surface pockets centered around X and Y. The iron-based superconductors are such systems, although the Fermiology can be more complicated, i.e. several hole and electron pockets are possible and this depends on the particular compound. In the following table, we assume one hole pocket and two electron pockets e_1 and e_2 around X and Y. The table can easily be extend. We make the following ansatz, so that the self-consistent gap equations (2.45) collapse to very few terms, because the higher harmonics are orthogonal to the leading harmonics that appear in the Ansatz above and the corresponding integral over angle φ vanishes identically. Thus, the self-consistent gap equations are an eigenvalue problem at the superconducting transition temperature $T = T_c$. It can be easily written in matrix form. At this point, one needs to determine the eigenvalues and eigenvectors for these linearized gap equations given by (2.45) and (2.46). Ideally, at least one of the eigenvalues is $\lambda = 1$, which means that the self-consistent equations are consistent, i.e. the manually introduced eigenvalue λ becomes obsolete. However, it does not have to be exactly $\lambda = 1$. In general, the largest eigenvalue will be realized by the system cooling down the temperature further and further. The corresponding eigenvectors then determines the relatives phases of the superconducting gaps, because every entry correspond to a Fermi surface pocket. There is one important caveat however. The superconducting state with the largest eigenvalue only realizes if the eigenvalue is positive. Otherwise, there is no chance to solve the self-consistent gap equations. This is due to the fact that the interaction potentials arise from the repulsive Coulomb interactions. In case of a conventional superconductor, the interaction potential or pairing vertex stems from phonon interaction and is attractive, thus it compensates the '$-$' sign on the right-hand side of the self-consistent equation. As was already pointed out, this sign change can be realized instead in case of an $s^{\pm}$-state, where the sign is introduced by a relative phase between the gaps of the hole and electron Fermi surface pockets instead. Another possibility is that the gap along the electron pocket acquires a strong momentum dependence, i.e. component $\bar{\Delta}_e$ is large, such that the gap has nodes at certain angles φ and changes its sign as well. The most promising situation however is the $s^{\pm}$-state.

If the momentum dependence is neglected everywhere for a moment, a quick analysis shows for a system with two degenerate hole Fermi surface pockets and two electron pockets around X and Y, the self-consistent gap equations in matrix form are

$$\lambda \begin{pmatrix} \Delta_{\mathrm{h}} \\ \Delta_{\mathrm{e}} \end{pmatrix} = - \begin{pmatrix} \Gamma_{\mathrm{hh}} & \Gamma_{\mathrm{he}} \\ \Gamma_{\mathrm{he}} & \Gamma_{\mathrm{ee}} \end{pmatrix} \begin{pmatrix} 2\Delta_{\mathrm{h}} \\ 2\Delta_{\mathrm{e}} \end{pmatrix}. \tag{2.54}$$

In this case, the eigenvalue λ is only positive if $\Gamma_{\mathrm{hh}}\Gamma_{\mathrm{ee}} < \Gamma_{\mathrm{he}}^2$. In words, this means that the inter-pocket scattering u_3 has to be larger than the intra-pocket scattering channels u_4 and u_5. In this context, we will discuss the role of spin-fluctuations in the formation of Cooper-pairs. However, before we proceed, we will motivate the necessity of spin-fluctuations for a minimal model that is tailored to capture the most basic properties of the iron-based superconductors.

2.5.2 *Superconductivity with the Bare Interactions*

Imagine the following system with one hole pocket, that has an alternating orbital content and is centered around Γ,

$$\alpha^\dagger = \cos(\varphi) d_{xz}^\dagger + \sin(\varphi) d_{yz}^\dagger, \tag{2.55}$$

and two electron pockets e_1 and e_2, which are centered around X and Y, and are dominated by either yz or xz orbital,

$$\beta_1^\dagger = d_{yz}^\dagger, \qquad \beta_2^\dagger = d_{xz}^\dagger. \tag{2.56}$$

We dropped the momentum and spin index for brevity. Note that the lobes of xz-orbital are usually elongated along the x-direction in real space, so that it is elongated along the k_y-direction in the momentum space and contributes mainly at the Y point instead of the X point. The model depicted above actually corresponds to the family of orbital-projected models. A good discussion of low-energy effective models is presented in Fernandes and Chubukov [25]. The Coulomb interactions can be written in terms of band operators rather quickly and the mean-field approach yields momentum dependent interaction potentials for the intra- and inter-pocket scattering terms. The momentum dependence will of course be a consequence of the coefficients from the hole pocket, where the orbital content is alternating. We approximate intra- and inter- orbital Coulomb repulsion to be roughly the same, $U = U'$, and drop contributions from J and J'. We can then determine the coefficients A^{mn}, B^{mn} and C^{mn} from the leading angular harmonics approximation. The resulting interactions potentials are given below, where we separated the s-wave and d-wave parts, which is possible because the different gap symmetries do not mix, i.e. are orthogonal, as was pointed out above. We defined

$$\mathcal{H}_{\text{int}}^{\text{orb}} = \sum_{i\nu\sigma} \frac{U}{2} d_{i\nu\sigma}^{\dagger} d_{i\nu\bar{\sigma}}^{\dagger} d_{i\nu\bar{\sigma}} d_{i\nu\sigma} + \sum_{i\nu\neq\mu\sigma\sigma'} \frac{U'}{2} d_{i\nu\sigma}^{\dagger} d_{i\mu\sigma'}^{\dagger} d_{i\mu\sigma'} d_{i\nu\sigma}$$
$$+ \sum_{i\nu\neq\mu\sigma\sigma'} \frac{J}{2} d_{i\nu\sigma}^{\dagger} d_{i\mu\sigma'}^{\dagger} d_{i\nu\sigma'} d_{i\mu\sigma} + \sum_{i\nu\sigma} \frac{J'}{2} d_{i\nu\sigma}^{\dagger} d_{i\nu\bar{\sigma}}^{\dagger} d_{i\mu\bar{\sigma}} d_{i\mu\sigma}. \tag{2.57}$$

and get

$$\text{s-wave}: \quad \Gamma_{\text{hh}}(\varphi,\varphi') = \Gamma_{\text{e}_j\text{e}_{j'}}(\varphi,\varphi') = \frac{U}{2}, \tag{2.58}$$

$$\text{d}_{x^2-y^2}\text{-wave}: \quad \Gamma_{\text{hh}}(\varphi,\varphi') = \frac{U}{2}\cos 2\varphi \cos 2\varphi', \tag{2.59}$$
$$\Gamma_{\text{he}_{1,2}}(\varphi,\varphi') = \mp\frac{U}{2}\cos 2\varphi,$$
$$\Gamma_{\text{e}_1\text{e}_1}(\varphi,\varphi') = \frac{U}{2} = \Gamma_{\text{e}_2\text{e}_2}(\varphi,\varphi'),$$
$$\Gamma_{\text{e}_1\text{e}_2}(\varphi,\varphi') = -\frac{U}{2} = \Gamma_{\text{e}_2\text{e}_1}(\varphi,\varphi'),$$

$$\text{d}_{xy}\text{-wave}: \quad \Gamma_{\text{hh}}(\varphi,\varphi') = \frac{U}{2}\sin 2\varphi \sin 2\varphi', \tag{2.60}$$
$$\Gamma_{\text{e}_j\text{e}_{j'}}(\varphi,\varphi') = 0,$$

where $j = 1, 2$ corresponds to e_1 and e_2. One can now use the s-wave and d-wave ansatz from Table 2.4 and evaluate the linearized gap equations (2.45) and (2.46). Below, the integration over angle φ' has already been performed. The resulting eigenvalue problem for the three different gap symmetries is

$$\begin{aligned} \text{s-wave}: \quad & \lambda \begin{pmatrix} \Delta_\text{h} \\ \Delta_\text{e} \end{pmatrix} = -\frac{U}{2} \begin{pmatrix} 1 & 2 \\ 1 & 2 \end{pmatrix} \begin{pmatrix} \Delta_\text{h} \\ \Delta_\text{e} \end{pmatrix}, \\ \text{d}_{x^2-y^2}\text{-wave}: \quad & \lambda \begin{pmatrix} \Delta_\text{h} \\ \Delta_\text{e} \end{pmatrix} = -\frac{U}{4} \begin{pmatrix} 1 & -4 \\ -4 & 1 \end{pmatrix} \begin{pmatrix} \Delta_\text{h} \\ \Delta_\text{e} \end{pmatrix}, \\ \text{d}_{xy}\text{-wave}: \quad & \lambda \begin{pmatrix} \Delta_\text{h} \\ \Delta_\text{e} \end{pmatrix} = -\frac{U}{2} \begin{pmatrix} 1 & 0 \\ 0 & 0 \end{pmatrix} \begin{pmatrix} \Delta_\text{h} \\ \Delta_\text{e} \end{pmatrix}. \end{aligned} \tag{2.61}$$

A positive eigenvalue λ appear $\Gamma_{\text{hh}}\Gamma_{\text{ee}} < \Gamma_{\text{he}}^2$ holds, which implies stronger inter-pocket scattering between hole and electron pockets. This as we see does not occur for the bare interactions (at least for the leading harmonics of the superconducting gaps). In the next section, we discuss how spin-fluctuations can indeed tip the system in the right direction, such that an $\text{s}^{\pm}$-state is formed where $\Delta_\text{h} \approx -\Delta_\text{e}$. Nevertheless at the level of the bare interactions the non-phononic mechanism of superconductivity is possible. However, it requires to include the higher harmonics in the corresponding irreducible representations, see for further review [50].

The situation becomes particularly simple if there is mechanism, which generally enhances the interband repulsion over the intraband ones. For example, for a

Table 2.4 Leading angular harmonics approximation of gap functions along the contour lines of the hole and electron Fermi surface pockets. The letter h denotes a hole pocket centered around Γ while e_1 and e_2 correspond to the electron pockets centered around X and Y respectively

	s-wave	$d_{x^2-y^2}$-wave	d_{xy}-wave
h	Δ_h	$\Delta_h \cos 2\varphi$	$\Delta_h \sin 2\varphi$
e_1	$\Delta_e + \bar{\Delta}_e \cos 2\varphi$	$\Delta_e + \bar{\Delta}_e \cos 2\varphi$	$\Delta_e \sin 2\varphi$
e_2	$\Delta_e - \bar{\Delta}_e \cos 2\varphi$	$-\Delta_e + \bar{\Delta}_e \cos 2\varphi$	$\Delta_e \sin 2\varphi$

multi-orbital system (orbital indexes p, q, s and t), the following tensor describes the general pairing vertex,

$$\Gamma_{st}^{pq}(\mathbf{k}, \mathbf{k}', \omega) = \left[\frac{3}{2}\hat{U}\chi^{RPA}(\mathbf{k}-\mathbf{k}', \omega)\hat{U} + \frac{1}{2}\hat{U} - \frac{1}{2}\hat{U}^c\chi_c^{RPA}(\mathbf{k}-\mathbf{k}', \omega)\hat{U}^c + \frac{1}{2}\hat{U}\right]_{st}^{pq}. \quad (2.62)$$

The interaction matrices $\hat{U}$ defined in (2.82) and $\hat{U}^c$, defined in Graser et al. [28] stem from the different Coulomb interactions. The spin susceptibility χ^{RPA} is defined diagrammatically in (2.69). Obviously, the inter-pocket scattering Γ_{he}, which includes $\mathbf{k} - \mathbf{k}' = \mathbf{Q}_{AF}$, is enhanced by the spin-susceptibility within Random Phase approximation due to the nesting of electron and hole bands. Hence, spin-fluctuations are considered to be sufficient to obtain $\lambda > 0$ and $\Gamma_{hh}\Gamma_{ee} < \Gamma_{he}^2$ in the simple model described in the previous section. However, if the spin-fluctuations are weak and the system has several hole and electron Fermi surface pockets, the conventional $s^{\pm}$-wave state, where the gaps change sign between hole and electron pockets, is not the only possible solution. For example in the case of the iron-based compound LiFeAs, there are three hole pockets with different orbital character. Two of these hole pockets are mainly $d_{xz,yz}$ orbitals, while the third hole pocket mainly stems from the d_{xy} orbital, like the electron pockets. As a consequence, LiFeAs has two almost distinct sets of Fermi surface pocket, which have a different orbital character, so that only inter-orbital Coulomb repulsion can contribute to inter-pocket scattering in the band basis. Hence, the two sets are less coupled than the pockets are coupled to pockets of the same set, and can be approximated as two almost separated multi-band superconductors. The implication for the linearized gap equations is that they also split into too decoupled subsets, which can in general be solved independently. This introduces an additional degree of freedom to the system. We sketch the possible solutions, under the assumption that inter-pocket scattering within both subsets is slightly enhanced compared to intra-pocket interaction (Fig. 2.9). In the following phase diagram, the effect of spin-fluctuations that enhance inter-pocket scattering between hole and electron pockets, is introduced manually by increasing the interactions $\Gamma_{h_i e_j}$ with regard to their bare value as derived from the projection of the bare Coulomb interactions on the various band channels that give rise to superconductivity. Note that the parameter of the system can in general be tuned, for example by applying pressure, so that a transition between the various s-wave states is possible as was sketched in Fig. 2.11b. In this case, at least one gap on a Fermi surface pocket

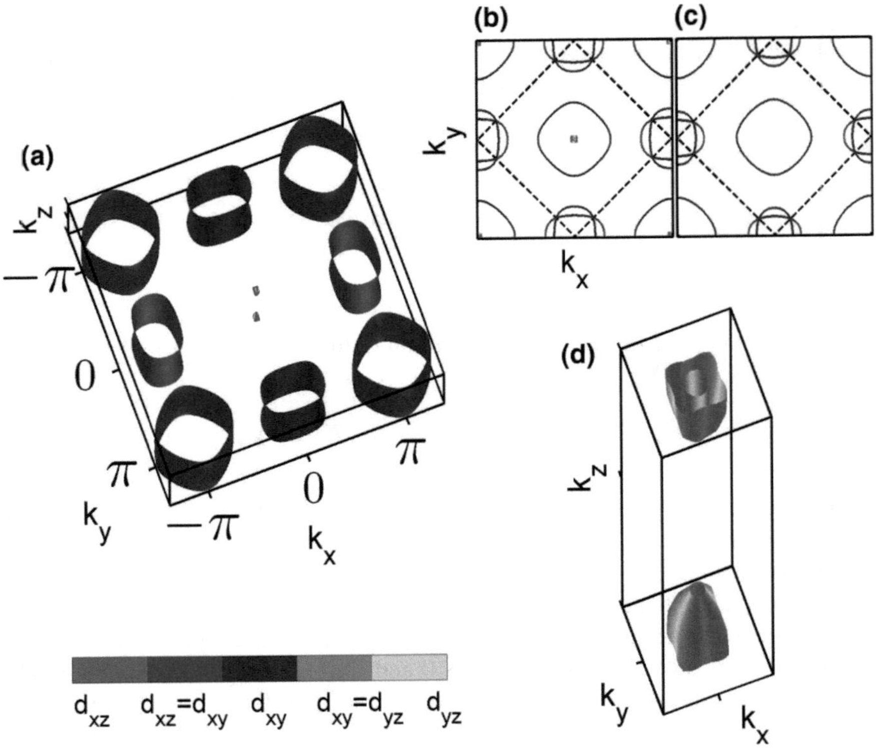

Fig. 2.9 The Fermi surface of the iron-based compound LiFeAs within tight-binding approximation that was adjusted to fit angle-resolved photoemission experiments and is defined here [74]. **a** Three-dimensional Fermi surface pockets. **b** The two small hole pockets are only present at $k_z = \pi$. At $k_z = 0$, they lie just below the Fermi level (**c**). These two hole pockets differ from the large hole pocket that is centered around M and the elliptic hole pockets, because they stem mainly from the $d_{xz,yz}$-orbitals (**d**)

has to vanish at the boundary of two different states. However, it might be energetically more favorable is non-trivial phase is acquired instead, so that magnitude of the gap is always non-zero. Such a state breaks the time-reversal symmetry and was discussed for example in [48]. Moreover, such a state is often dubbed s+is state, which emphasizes the superposition of two distinct s-wave states. A mixture with a d-wave state is also in general possible and is called s+id state. However, one finds that the $d_{x^2-y^2}$-state is a solution of the linearized gap equations, but it always looses against s-wave (see Fig. 2.11), i.e. the corresponding eigenvalue is smaller so that it can only realize at a smaller temperature if s-wave is suppressed artificially.

To summarize, the orbital effects on superconductivity are mostly pronounced in the situaions when the orbital character of the Fermi surface varies between different bands. In such a case the superconducting gap can be characterized not only as sign

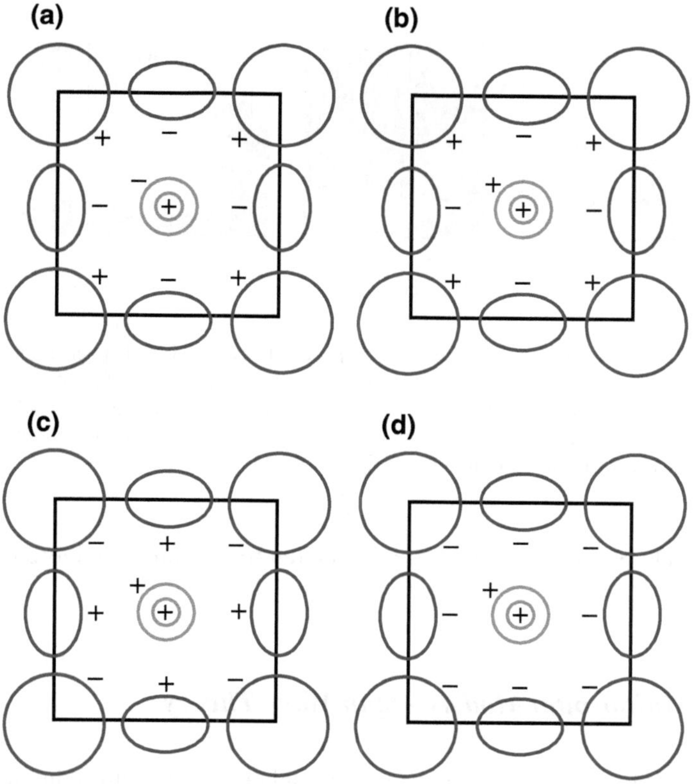

Fig. 2.10 Sketch of the Fermi surface of the stoichiometric superconductor LiFeAs and the relative sign of the corresponding gaps. Cooper-pairing from repulsive interaction can be overcome if a relative phase between gaps of the different Fermi surface pockets is introduced and the conventional $s^{\pm}$-state has a global sign change between circular hole and elliptic electron pockets as shown in B. However, the phase diagram shows a rich structure because mainly four states are supported by weak spin-fluctuations

changing but also orbital-antiphasing, which might be the case for some of the iron-based superconductors like LiFeAs. Nevertheless, for majority of the systems where antiferromagnetism is also present the superconducting gap can be characterized as a global $s^{\pm}$ with sign-changing gap between electron and hole pockets.

2.6 Spin Waves in Itinerant Multiorbital Systems

Now we also would like to give further technical details in how to calculate the spin excitations below T_N in itinerant multiorbital systems. This requires a generalization

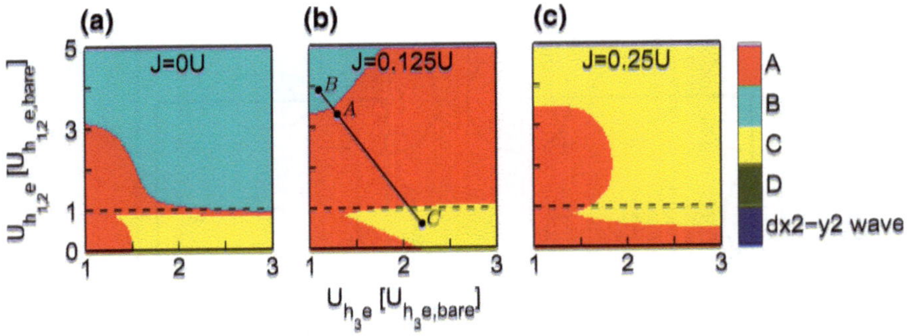

Fig. 2.11 The bare interaction potentials $U_{h_{1,2}e}$ involve the two small hole pockets and the elliptic electron pockets of LiFeAs, while U_{h_3e} involves the large hole pocket. The effect of spin-fluctuations at wave-vector $\mathbf{Q}_{AF}$ is taken into account by renormalizing and increasing inter-pocket scattering for those pockets that are nested at this vector. In case of intermediate values of Hund J (**b**), the unconventional s-wave state B is dominant (Fig. 2.10)

of mean-field plus random phase approximation formalism to multiorbital systems (MOS).

2.6.1 Multiorbital Models - Spin Wave Theory

The main ideas are similar as for the single band model, but with additional complications due to orbital indices. In the following we follow closely [38]. We define the spin operator as

$$\mathbf{S}_s(\mathbf{q}) = \sum_{\mathbf{k},\sigma,\sigma'} \psi^\dagger_{s\sigma}(\mathbf{k}+\mathbf{q})\frac{\sigma_{\sigma\sigma'}}{2}\psi_{s\sigma'}(\mathbf{k}), \tag{2.63}$$

where $\psi^\dagger_{s\sigma}$ is the creation operator of the orbital s with spin σ in the MOS. The transverse spin susceptibility in the paramagnetic state can be calculated from the spin-spin correlation function in Matsubara frequency space:

$$\chi^{\pm}_{st}(\mathbf{q}, i\Omega) = \int_0^\beta d\tau e^{i\Omega\tau}\left\langle T_\tau S^+_s(\mathbf{q},\tau)S^-_t(-\mathbf{q},0)\right\rangle. \tag{2.64}$$

In MOS it is convenient to introduce a dynamical spin susceptibility tensor in a more general way using four orbital indices [1, 34]

$$\chi^{\pm}_{spqt}(\mathbf{q}, i\Omega) = \int_0^\beta d\tau e^{i\Omega\tau} \sum_{\mathbf{k},\mathbf{k}',\gamma,\delta,\gamma',\delta'} \sigma^+_{\gamma\delta}\sigma^-_{\gamma'\delta'} \tag{2.65}$$

$$\left\langle T_\tau \psi^\dagger_{p\gamma}(\mathbf{k},\tau)\psi_{q\delta}(\mathbf{k}+\mathbf{q},\tau)\psi^\dagger_{t\gamma'}(\mathbf{k}',0)\psi_{s\delta'}(\mathbf{k}'-\mathbf{q},0)\right\rangle. \tag{2.66}$$

With the Green's function (GF)

$$G^{\sigma}_{st}(\mathbf{k}, i\omega_n) = -\int_0^{\beta} d\tau e^{i\omega\tau} \langle T_\tau \psi_{s\sigma}(\mathbf{k}, \tau)\psi^{\dagger}_{t\sigma}(\mathbf{k}, 0)\rangle \tag{2.67}$$

the susceptibility matrix can be written as

$$\chi^{\pm}_{spqt}(\mathbf{q}, i\Omega) = -\frac{1}{N\beta}\sum_{\mathbf{k}, i\omega_n} G^{\uparrow}_{sp}(\mathbf{k}, i\omega_n)G^{\downarrow}_{qt}(\mathbf{k}+\mathbf{q}, i\omega_n + i\Omega) \tag{2.68}$$

The susceptibility in the paramagnetic phase within the random phase approximation can be written in a matrix form $\hat{\chi}$, see for example Kemper et al. [34]

$$\chi^{RPA}_{bsta} = \chi^{0}_{bsta} + \chi^{RPA}_{b'sta'} U_{c'b'a'd'} \chi^{0}_{bc'd'a} \quad , \tag{2.69}$$

and the solution of the RPA in matrix form is straightforward.

$$[\hat{\chi}]^{RPA} = \hat{\chi}^0(1 - \hat{U}\hat{\chi}^0)^{-1} \tag{2.70}$$

The physical susceptibility which is measured in experiment is then obtained

$$\chi^{RPA}_{phys}(\mathbf{q}, \omega) = \frac{1}{2}\sum_{st} \chi^{RPA}_{sstt}(\mathbf{q}, \omega). \tag{2.71}$$

However, important information is also contained in the individual components of the susceptibility tensor χ_{spqt} which can be used to distinguish between predominantly intra- or interorbital fluctuations. The random phase approximation in diagrammatic form is illustrated in Fig. 2.12.

Generally, for MOS, the kinetic or tight-binding part H_0 of the Hamiltonian $H = H_0 + H_{int}$, including the intra- and inter-orbital hoppings, can be diagonalized via a unitary transformations which gives the eigenergies $E_\nu(\mathbf{k})$ of the system. The operators in the orbital and the band basis are connected via the coefficients of the

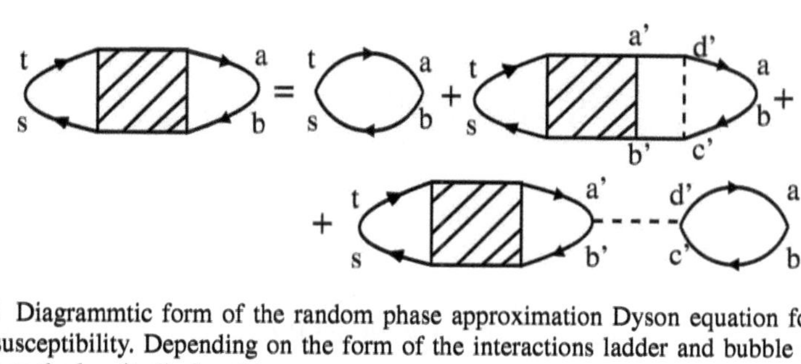

Fig. 2.12 Diagrammtic form of the random phase approximation Dyson equation for the multiorbital susceptibility. Depending on the form of the interactions ladder and bubble diagramms contribute to the longitudinal and transverse component

unitary transformations $a_\nu^s(\mathbf{k},\sigma)$

$$\psi_{s\sigma}(\mathbf{k}) = \sum_\nu a_\nu^s(\mathbf{k},\sigma)\gamma_{\nu\sigma}(\mathbf{k}) \tag{2.72}$$

Without interactions the GFs are easily obtained as

$$G_{st}^\sigma(\mathbf{k}, i\omega_n) = \sum_\nu \frac{a_\nu^s(\mathbf{k},\sigma)a_\nu^{t*}(\mathbf{k},\sigma)}{i\omega - E_\nu(\mathbf{k})} \tag{2.73}$$

and the transverse spin susceptibility becomes:

$$\chi_{spqt}^{\pm}(\mathbf{q}, i\Omega) = -\frac{1}{N\beta}\sum_{\mathbf{k},i\omega,\nu,\mu} \tag{2.74}$$

$$\frac{a_\mu^s(\mathbf{k},\uparrow)a_\mu^{p*}(\mathbf{k},\uparrow)a_\nu^q(\mathbf{k}+\mathbf{q},\downarrow)a_\nu^{t*}(\mathbf{k}+\mathbf{q},\downarrow)}{\left[i\omega - E_\mu(\mathbf{k})\right]\left[i\omega + i\Omega - E_\nu(\mathbf{k}+\mathbf{q})\right]}. \tag{2.75}$$

The Matsubara sum can be performed via a standard contour integration and by analytic continuation $i\Omega \to \Omega + i\Gamma$. Observe that for large Hamiltonian matrices the coefficients a_ν have to be calculated numerically by diagonalizing the Hamiltonian matrix written in **k**-space.

As described above in a mean-field treatment of the spin density wave phase, the interaction part of the hamiltonian H_{int} is decoupled with respect to a groundstate with a finite magnetization at wave vector **Q**, e.g. $(\pi, 0)$ or $(0, \pi)$ for iron-based superconductors. Here, we assume that only one of them is present below T_N and also negelect its momentum dependence due to orbital matrix elements. The order parameters read

$$\Delta_{\nu\nu'} = \sum_{\mathbf{k}} \mathrm{sgn}(\sigma)\left\langle \psi_{\nu\sigma}^\dagger(\mathbf{k})\psi_{\nu'\sigma}(\mathbf{k}+\mathbf{Q})\right\rangle. \tag{2.76}$$

In the magnetic state, SU(2) spin symmetry is broken which yields $\chi^{\pm} \neq \frac{1}{2}\chi^{zz}$. In addition, the antiferromagnetic state also breaks translational symmetry and the mean-field Hamiltonian in the reduced Brillouin zone takes the form

$$H_{SDW} = {\sum_{\mathbf{k}}}' \hat{\psi}_\sigma^\dagger(\mathbf{k}) \begin{pmatrix} H_0(\mathbf{k}) & \sigma M \\ \sigma M & H_0(\mathbf{k}+\mathbf{Q}) \end{pmatrix} \hat{\psi}_\sigma(\mathbf{k}) \tag{2.77}$$

where we introduce the new spinor

$$\hat{\psi}_\sigma(\mathbf{k}) = \begin{pmatrix} \psi_\sigma(\mathbf{k}) \\ \psi_\sigma(\mathbf{k}+\mathbf{Q}) \end{pmatrix}. \tag{2.78}$$

Note for MOS each component of the spinor may acquire a spinor structure in orbital space as well. Then the GF acquires off-diagonal elements in momentum space and becomes a matrix. In short-hand notation, we write only the orbital index with a tilde to indicate that the momentum is shifted by the antiferromagnetic wave vector. We thus obtain

$$\hat{G}_{sp} = \left\langle \begin{pmatrix} s \\ \tilde{s} \end{pmatrix} \begin{pmatrix} p^\dagger & \tilde{p}^\dagger \end{pmatrix} \right\rangle = \begin{pmatrix} G_{sp} & G_{s\tilde{p}} \\ G_{\tilde{s}p} & G_{\tilde{s}\tilde{p}} \end{pmatrix} \tag{2.79}$$

Analogously, the susceptibility matrix becomes a matrix in **q**-space [1, 36, 63]:

$$\hat{\chi}^{\pm}_{spqt} = \begin{pmatrix} \chi^{\pm}_{spqt}(\mathbf{q}, \mathbf{q}) & \chi^{\pm}_{spqt}(\mathbf{q}, \mathbf{q}+\mathbf{Q}) \\ \chi^{\pm}_{spqt}(\mathbf{q}+\mathbf{Q}, \mathbf{q}) & \chi^{\pm}_{spqt}(\mathbf{q}+\mathbf{Q}, \mathbf{q}+\mathbf{Q}) \end{pmatrix} \tag{2.80}$$

where the off-diagonal components, in short hand notation $\chi^{\pm}_{spqt}(\mathbf{q}, \mathbf{q}+\mathbf{Q}) = \left[G_{sp}G_{q\tilde{t}} + G_{s\tilde{p}}G_{\tilde{q}\tilde{t}} + G_{\tilde{s}\tilde{p}}G_{\tilde{q}t} + G_{\tilde{s}p}G_{qt}\right]$ are Umklapp susceptibilities as they connect spins with momentum shifted by the AFM wave vector, which is a reciprocal wave vector of the magnetic Brillouin zone. The detailed expressions can be found in a number of references [38, 39, 62].

The form of the random phase approximation in the spin density wave state remains the same, see (2.69), with the only extension that the interaction matrix also doubles in size with $\hat{U}$ matrices on the diagonal and zeros on the off-diagonal. To obtain the physical susceptibility we perform an random phase approximation with (2.80) and trace over the orbital indices of the first entry $\chi^{RPA}_{spqt}(\mathbf{q}, \mathbf{q})$ of the susceptibility matrix. The imaginary part is directly related to the cross section of INS measurements. We also recall that self-consistency enters via taking the same set of interaction parameters in the random phase approximation and the Hartree-Fock mean-field equations.

2.6.2 *Itinerant Frustration and Accidental Zero Modes*

Before computing spin waves for realistic parameters of iron-based superconductors we again come back to the idealized model of fully nested electron and hole pockets discussed in the previous section. At a mean-field level there is a large ground state degeneracy. In a localized model this would be reflected by additional soft modes in the spin wave spectra. It is interesting to check whether signs of frustration in the magnetic excitation spectrum also show up in the itinerant description. In Fig. 2.13 we show the transverse (upper panel) and longitudinal (lower panel) components of the imaginary part of the spin susceptibility for zero and finite ellipticities. Observe that the magnetic order is fixed to be along $(\pi, 0)$ wave vector. Consider first the idealized case of circular electron pockets, $\varepsilon = 0$ (left panels). In this case two out of three FS are completely gapped. In Fig. 2.13a we show the calculated imaginary part of the transverse component of the susceptibility, Im $\left[\hat{\chi}_{+-}(\mathbf{q}, \Omega)\right]_{RPA}$ along the

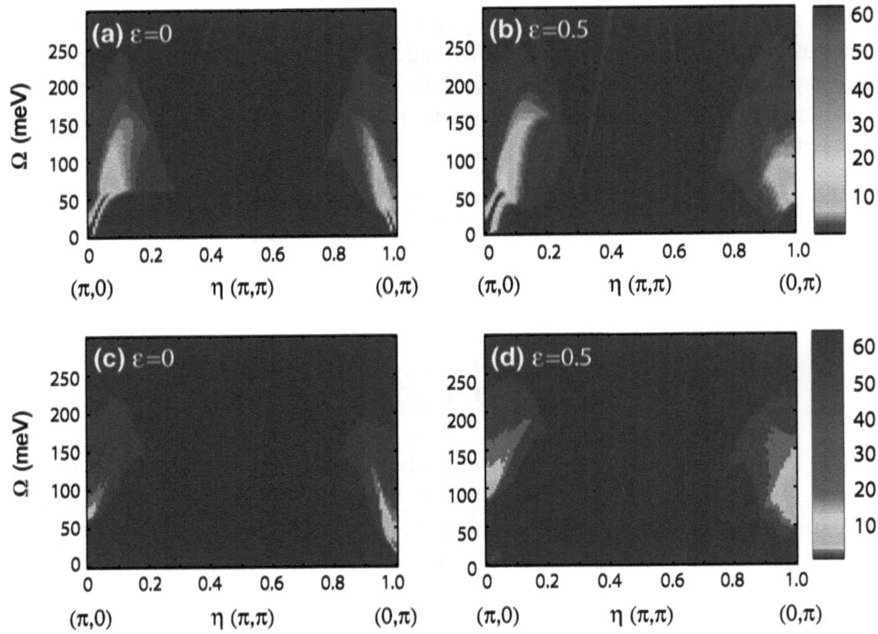

Fig. 2.13 Imaginary part of the transvers (*upper panel*) and longitudinal (*lower panel*) susceptibility along $(0, \pi) \to (\pi, 0)$ points of the Brillouin zone for two different ellipticities of the elctron pockets parameterized by ε. Taken from Knolle et al. [36]

direction $(0, \pi)$ to $(\pi, 0)$. Because the corresponding ground state is degenerate, the excitation spectrum has a Goldstone mode at the ordering momentum $\mathbf{Q}_1 = (\pi, 0)$ and another accidental gapless mode at $\mathbf{Q}_2 = (0, \pi)$. The excitations near the ordering momentum are propagating up to $\Omega_c(\mathbf{Q}_1) = 2\Delta_1 \sim 60$ meV which is expected because both pockets separated by $\mathbf{Q}_1$ are gapped. Excitations near $\mathbf{Q}_2$ are also propagating, but up to smaller energies, $\Omega_c(\mathbf{Q}_2) = \Delta_1$, that is consistent with the fact that only one of the two pockets separated by $\mathbf{Q}_2$ is gapped. At higher energies spin-waves enter the continuum and damped Stoner-like excitations, however, still with well defined peaks. The difference between $\Omega_c(\mathbf{Q}_1)$ and $\Omega_c(\mathbf{Q}_2)$ is also seen in Fig. 2.13c where the longitudinal component of the susceptibility is plotted. As expected, $\mathrm{Im}\big[\hat{\chi}_{+-}(\mathbf{q}, \Omega)\big]_{RPA}$ vanishes below $\Omega_c(\mathbf{Q})$.

For finite $\varepsilon = 0.5$ the results are presented in panels (b) and (d). There are two key effects introduced by ellipticity. First, the degeneracy is now lifted [19] and the transverse excitations near $\mathbf{Q}_2$ acquire a finite gap clearly visible in Fig. 2.13b. Second, spin-wave excitations near $\mathbf{Q}_1$ have a finite Landau damping down to $\Omega = 0$ because the spin density wave order no longer completely gaps the hole and electron FSs separated by $\mathbf{Q}_1$. At the same time, spin-wave excitations are still clearly visible at low energies and do not become overdamped. The reason is that spin density wave coherence factors suppress the scattering between electron and hole like pockets involved in the formation of magnetic order and reduce the Landau damping around the ordering momentum $\mathbf{Q}_1$ from $\Omega/|\mathbf{q}-\mathbf{Q}_1|$ to $\Omega|\mathbf{q}-\mathbf{Q}_1|$ what for linear dispersion is of the same order as Ω^2 and $(\mathbf{q}-\mathbf{Q}_1)^2$. Once Ω exceeds $\Omega_c(\mathbf{Q}_2)$, the coherence factors no longer screen Landau damping, and spin-waves become Stoner-like, overdamped excitations. The scattering between between the pockets not involeved in magnetism is not suppressed by the spin density wave coherence factors because $\mathbf{Q}_2$ is not the ordering momentum, therefore, Landau damping makes the transverse excitations near $\mathbf{Q}_2$ overdamped immediately above the gap. Note, that there is also a tendency towards incommensuration near $\mathbf{Q}_2$. Longitudinal excitations are still gapped up to $\Omega_c(\mathbf{Q})$ and become Stoner-like at higher energies.

Overall, it is interesting to observe that in our itinerant description the groundstate degeneracy for the fully nested situation which arises due to frustration between the two wave vectors $\mathbf{Q}_1$ and $\mathbf{Q}_2$ shows up in the magnetic excitation spectrum as an additional soft mode.

2.6.3 Two Orbital Model: Orbital Versus Excitonic Scenario

So far we have not discussed the role of orbitals in the formation of the antiferromagnetic order. At the same time, there is no doubt that the orbitals do play an important role. For example they introduce a significant angular dependence for the intraband and interband interactions resulting in a change of important details in the superconducting and antiferromagnetic states. It was argued that the originally nodeless s^{+-}-wave superconducting state may acquire nodes on the electron pockets. Maiti et al. [47], Thomale et al. [70], Kemper et al. [34] For the antiferromagnetic state it was predicted that iron-based superconductors show nodal elementary excitations with a linear Dirac spectrum at low energies as a result of orbital effects [54, 60, 76]. In addition, due to the fact that the antiferromagnetic order is either $(\pi, 0)$ or $(0, \pi)$, the magnetic anisotropy induces orbital anisotropies which in turn affect the electronic properties along a and b directions [15, 67, 72]. They may also introduce novel features such as ferro-orbital or antiferro-orbital order [41, 46] associated with intra and inter-orbital electron-hole pairing. Therefore, band and orbital lattice descriptions are not necessarily equivalent even in the limit of interactions being much smaller than the bandwidth. Furthermore, nesting features do not play an essential role in the formation of antiferromagnetic order starting from the orbital description.

The susceptibility matrix for the interacting part of the Hamiltonian previously defined in (2.16), with labels 1 for xz and 2 for yz orbitals, reads

$$\hat{\chi} = \begin{bmatrix} \chi_{1111} & \chi_{2111} & \chi_{1112} & \chi_{2112} \\ \chi_{1211} & \chi_{2211} & \chi_{1212} & \chi_{2212} \\ \chi_{1121} & \chi_{2121} & \chi_{1122} & \chi_{2122} \\ \chi_{1221} & \chi_{2221} & \chi_{1222} & \chi_{2222} \end{bmatrix} \tag{2.81}$$

and the interaction matrix for the transverse component of the spin susceptibility is given by:

$$\hat{U} = \begin{bmatrix} U & 0 & 0 & J \\ 0 & J & U-2J & 0 \\ 0 & U-2J & J & 0 \\ J & 0 & 0 & U \end{bmatrix}. \tag{2.82}$$

Following the procedure described above it is straightforward to compute the physical susceptibility in the paramagnetic state. The response is governed by antiferromagnetic fluctuations centered at $(\pi, 0)$ and $(0, \pi)$ resulting from scattering that connects the hole pocket around (0, 0) and the two electron pockets at the Brillouin zone boundaries.

In the next step we introduce self-consistent mean-field averages associated with the antiferromagnetic state. We consider two scenarios. In the first case, which we denote as *orbital*, one introduces the magnetic state in the orbital basis using the local interactions. The mean-field phase diagram of the two orbital model was studied by several groups [41, 49, 60] and the main conclusion is that antiferromagnetic ordering involves only very weakly interorbital components. Thus we assume in the following that the mean-field antiferromagnetic order involves only states within the same orbital, separated by $\mathbf{Q}_{AF}$ ($\Delta_{\nu\nu'} = \delta_{\nu,\nu'}\Delta_\nu$), and decouple the on-site local interaction, (2.16), with respect to it. In addition, we set $\mathbf{Q}_{AF} = (0, \pi)$ and do not consider the problem of its selection out of the manifold of degenerate magnetic states as it was discussed in the first section. We obtain two coupled gap equations

$$M_x = -\left[U\Delta_{xz} + J\Delta_{yz}\right] \tag{2.83}$$
$$M_y = -\left[U\Delta_{yz} + J\Delta_{xz}\right] \tag{2.84}$$

and the Hamiltonian matrix becomes

$$\hat{h}^{Orbital} = \begin{pmatrix} \varepsilon_x & \varepsilon_{xy} & \sigma M_x & 0 \\ \varepsilon_{xy} & \varepsilon_y & 0 & \sigma M_y \\ \sigma M_x & 0 & \tilde{\varepsilon}_x & \tilde{\varepsilon}_{xy} \\ 0 & \sigma M_y & \tilde{\varepsilon}_{xy} & \tilde{\varepsilon}_y \end{pmatrix} \tag{2.85}$$

where the tilde again means that the momentum is shifted by $\mathbf{Q}_{AF}$ The matrix can be inverted and with the new eigenvectors one can solve the coupled gap equations and carry out the spin wave calculations as outlined above.

The second, *excitonic*, scenario is built on the belief that in the iron-based superconductors, the antiferromagnetic instability, at least partially, is connected to nesting of electron and hole bands [9, 53]. This scenario we largely explored in the previous subsection. Within this scenario the tight-binding hamiltonian, (2.14), is first diagonalized to obtain the band structure. One further assumes, that the bands are the relevant degrees of freedom and ignores the role of the underlying orbital structure. The interactions are then also re-written in the band representation. Moreover, within the RG flow they are strongly renormalized, to such an extent that their values at the Fermi energy are not the same as the initial ones determined by the local on-site interactions [9]. This allows us to approximate the band interactions as Fermi-liquid like density-density type interactions given in (1.7), which at the lowest order are angular independent. By decoupling the band interaction with respect to Δ_{12} (we assume now only one wave vector for the order parameter) one finds a single gap equation

$$M = -\frac{u_1 + u_3}{2}\Delta_{12} \tag{2.86}$$

and the Hamiltonian matrix becomes

$$\hat{h}^{Band} = \begin{pmatrix} E_1 & 0 & 0 & \sigma M \\ 0 & E_2 & \sigma M & 0 \\ 0 & \sigma M & \tilde{E}_1 & 0 \\ \sigma M & 0 & 0 & \tilde{E}_2 \end{pmatrix}. \tag{2.87}$$

When comparing the orbital and the excitonic scenario one realizes that for small magnetic moments and low energies the spin excitations look very similar. Knolle et al. [38] The Fermi surface has tiny pockets in both cases, however, in the excitonic scenario they are due to imperfect nesting and disappear quickly for larger moments, whereas, in the orbital picture the pockets are coming from Dirac cones away from the Fermi energy and are thus protected even at larger interactions [60]. In Fig. 2.14 we show the transverse susceptibility for a magnetic moment of $m = 0.38\mu_B$ in both scenarios along the route $(\pi, 0) \to (0, 0) \to (0, \pi) \to (\pi, \pi)$ of the first Brillouin zone. The peculiar feature of the spin excitations in the antiferromagnetic state in both scenarios are well-defined spin waves around the ordering wave vector. They disperse until they hit the particle-hole continuum and become overdamped.

As we pointed out in the previous subsection there is an $O(6)$ degeneracy of the magnetic ground state [even in the orbital model] which manifests itself in additional zero energy modes at the non-ordered momenta of $(\pi, 0)$ and (π, π) for larger magnetic moments (not shown, see Knolle et al. [38]). These modes, however, are again accidental in the sense that they are not stable with regard to perturbations such as the finite Landau damping introduced by remnant Fermi surfaces. This is why they are not seen for small magnetic moments. However, as soon as the FSs are com-

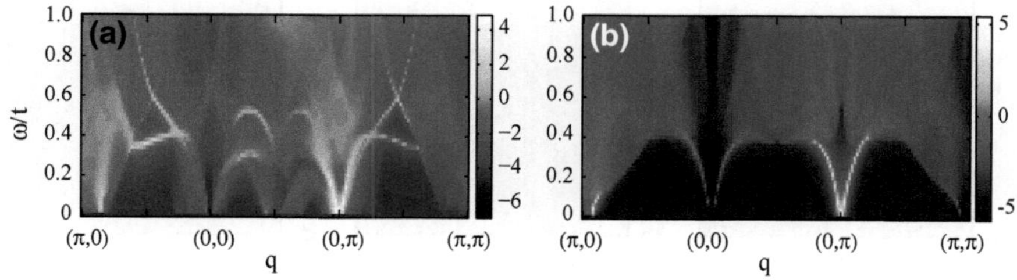

Fig. 2.14 Imaginary part of the transverse spin susceptibility in the orbital (**a**) and excitonic (**b**) scenarios for a magnetic moment of $m = 0.38\mu_B$. Taken from Knolle et al. [38]

pletely gapped they become visible. In the orbital scenario, however, the FSs will be gapped completely only for very large interactions for symmetry reasons and the particle-hole continuum remains gapless. Thus, the spin waves remain well-defined only around the ordering momentum. Away from it they are overdamped because the Dirac cones in the electronic excitation spectra lead to remnant FS pockets.

We show in Fig. 2.15 constant energy cuts for both scenarios on an absolute scale. In both cases the elliptical shaped ring around the ordering momentum (center of the plots) disperses at low energies until it enters the particle-hole continuum. Due to the orbital structure of the wave functions the response in the orbital scenario has a lot more structure which is supported by recent inelastic neutron scattering experiments [20].

A peculiarity of the spin waves within the orbital scenario is the relatively rich structure of the spin excitations away from the ordering momentum. Here the complication is that each orbital has its own magnetic gap which correspondingly affects the magnetic excitations within each orbital. It is possible to analyze separately the intra and inter-orbital components of the transverse spin susceptibility. Within the orbital scenario, the Goldstone modes originate from the intra-orbital *xz* and *yz* susceptibilities, while in the excitonic scenario the entire low energy excitations are in the inter-band susceptibility by construction. The antiferromagnetic magnetic order breaks the symmetry between *xz* and *yz* orbitals which leads to different behavior of the spin excitations away from the ordering momentum and at higher energies

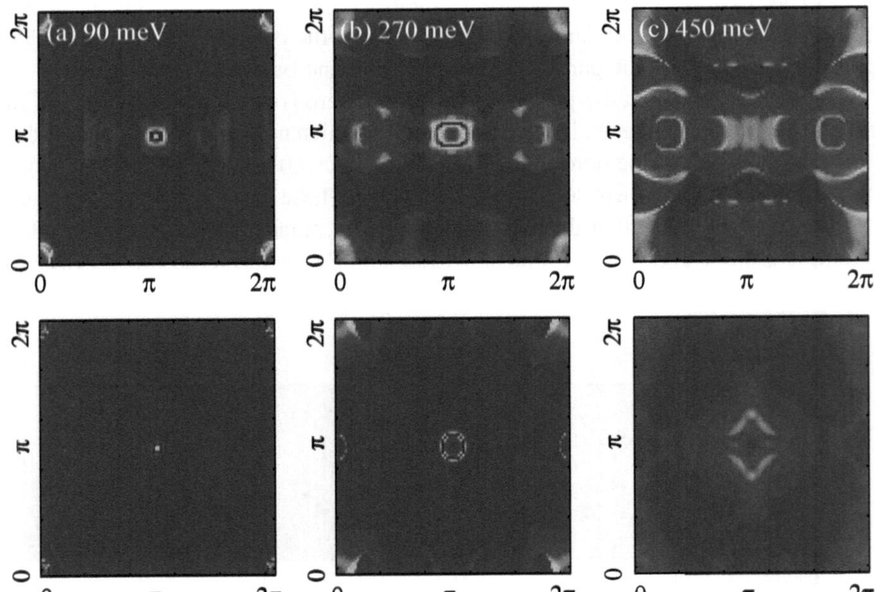

Fig. 2.15 Constant energy cuts of the imaginary part of the total transverse susceptibilities in the orbital (*upper*) and excitonic (*lower*) scenarios for a magnetic moment of $\mu = 0.38\mu_B$ and three different energies

in the corresponding intra-orbital susceptibilities. Given the fact that the spin waves show some orbital dependencies we believe that this feature is measurable in RIXS experiments [32, 38].

Another interesting feature of the spin excitations in the orbital scenario is that their spin wave branches can reappear out of the continuum at higher energies, e.g. when going from $(0, 0)$ to $(0, \pi)$ as shown in Fig. 2.14 Here, orbital coherence factors yielding the larger continuum (damping) are small at these higher energies, thus allowing the spin waves to reappear. As soon as the itinerant magnetism appears in MOS with multiple order parameters this reappearance of spin waves at higher energies away from the ordering momentum should be a general feature not restricted to iron-based superconductors. Spin excitations within the orbital scenario show a rich structure at energies which exceed well the onset of the particle-hole continuum. Note that at these energies the spin excitations should not any longer be interpreted as spin waves originating from breaking of the spin-rotational symmetry in the antiferromagnetic state but rather should be attributed to a Stoner continuum. These excitations are almost unchanged across the Neel temperature and as such can be well separated in the experimental data.

A natural next step would be to carry out spin wave calculations with more accurate multiorbital models. Kaneshita and Tohyama [31] have done this for a more realistic five orbital model and found good agreement with experiments. Ewings et al. [20] However, there are various models used in the literature with slightly varying parameters. Depending on the choice of model and dispersion input the results may actually differ. Brydon et al. [2] We also note that more recent calculations for the more realistic Hamiltonian are presented in Kovacic et al. [39], Scherer et al. [62].

2.6.4 Comparison to Experiments

After having introduced the general theory how to calculate spin waves in itinerant magnets with multiple orbitals (bands) contributing at the Fermi energy we presented results for a two orbital (band) model. In the following, we want to compare the itinerant description of magnetism in ferropnictides with actual experiments. As mentioned above the two orbital model can give qualitative insights especially to elucidate the role of orbital degrees of freedom. However, in addition to the wrong electronic struture in the unfolded zone it is not possible to fit the dispersion parameters of the model quantitatively to experiments such as angle-resolved photoemission [59]. Therefore, we will show the results for a phenomenological three band model which can be fitted to experiments [36].

Neutron scattering measurements on parent *FeAs* compounds revealed that magnetic order consists of ferromagnetic chains along one crystallographic direction in an *Fe* plane and antiferromagnetic chains along the other [13, 35], i.e., the system selects $\mathbf{Q}_{AF}$ to be either $\mathbf{Q}_1 = (\pi, 0)$ *or* $\mathbf{Q}_2 = (0, \pi)$. Sharp propagating spin-wave like excitations have been observed near the ordering momentum (e.g., $\mathbf{Q}_1$) up to energies of around 100 meV, with different velocities along the two crystallographic directions [16, 82]. At larger energies, excitations become overdamped [16].

Magnetism with $(\pi, 0)$ stripe order arises also in the $J_1 - J_2$ model of localized spins for $J_2 < 0.5J_1$ [5], and it was argued that such a model can successfully describe some of the experimental data on the spin wave spectra, although, an unusually large in-plane anisotropy of the antiferromagnetic exchange between nearest neighbor spins has to be assumed [82] in order to account for the anisotropic spin wave velocities. We have seen in the first section of this review that it is possible to understand the order from an itinerant perspective, which takes into account the peculiar electronic structure of the ferropnictides. The ellipticity of the electron pockets and interactions between them played a key role. The issue we address now is whether inelastic neutron scattering (INS) measurements of magnetic excitations can be explained within the itinerant approach, as well. This is a crucial test of the itinerant description of magnetism in ferropnictides.

In Fig. 2.16 we compare the calculations performed in Knolle et al. [36] with inelastic neutron scattering experiments. Measured excitations near $\mathbf{Q}_1$ are propagating spin-waves up to about 100 meV, with a finite but small damping. In the left panel of Fig. 2.16 we compare the dispersion with the experimental data [16, 82]. We emphasize that both the measured and the calculated spectra have different velocities along different crystallographic directions. In the itinerant theory, it is a consequence of the non-zero ellipticity *and* the fact that only one electron pocket is involved in the spin density wave formation. Remarkably, for $\varepsilon = 0.5$ ellipticity which agrees with the angle-resolved photoemission data the anisotropy of the spin wave velocities is the same as in the experimental INS data. Another verifiable theory prediction for this range is that the width of the spin-wave peak should scale as Ω^2. Additionally, the excitations are still visible even when the spin-waves enter the continuum. In the right panel of Fig. 2.16 we compare the calculated and the measured $Im\chi^{+-}(q, \Omega)$ along $(0, 0) \rightarrow (\pi, 0) \rightarrow (2\pi, 0)$ direction for different Ω. We see that the agreement is quite good even for higher frequencies.

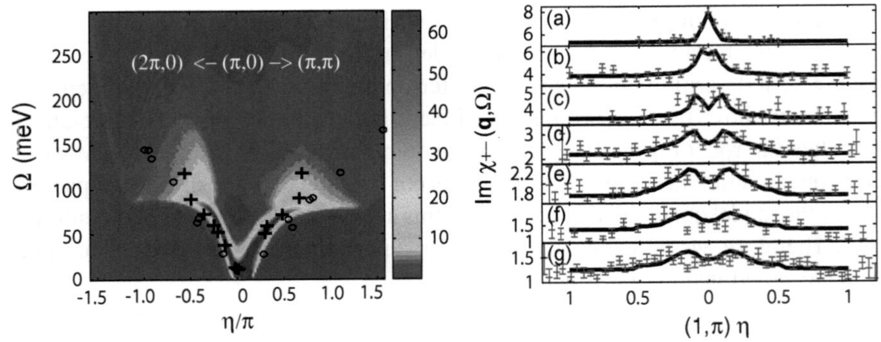

Fig. 2.16 Comparison of the calculated spin excitations to experimental data of Diallo et al. [16], Zhao et al. [82]. Taken from Knolle et al. [36]

Finally we would like to notice that further insight into the spin fluctuations in the paramagnetic state of the iron-based superconductors and at intermediate energies can be achieved within dynamical mean-field theory, which includes the local self-energy of the quasi-articles [80]. The remarkable agreement between theory and experiment has been achieved especially with regard to the contribution of the various orbitals to the spin fluctuations at various energy scales [42].

2.7 Discussion

In this chapter we have introduced the itinerant description of antiferromagnetism in iron-based superconductors. In contrast to theories built around purely localized spin model our theory, based on the itinerant description, allows for a coherent understanding of the full phase diagram of these materials as a function of doping, disorder and pressure. All the magnetic properties such as magnitude of the magnetic moment, selection of the order from the degenerate manifold of the possible ground states, orbital effects and finally the anisotropic spin waves can be well understood and connected to the basic electronic structure of these materials. A particular advantage is the possibility to compute charge and spin response functions in a consistent fashion within a single formalism and for the same set of parameters.

References

1. P.M.R. Brydon, C. Timm, Phys. Rev. B **80**, 174401 (2009); **79**, 180504(R) (2009)
2. P.M.R. Brydon, Maria Daghofer, Carsten Timm. J. Phys.: Condens. Matter **23**, 246001 (2011)
3. A. Cano, M. Civelli, I. Eremin, I. Paul, Phys. Rev. B **82**, 020408(R) (2010)
4. A. Carrington, Rep. Prog. Phys. **74**, 124507 (2011)
5. P. Chandra, P. Coleman, A.I. Larkin, Phys. Rev. Lett. **64**, 88 (1990)
6. J.-H. Chu, J.G. Analytis, K. De Greve, P.L. McMahon, Z. Islam, Y. Yamamoto, I.R. Fisher, Science **329**, 824 (2010)
7. T.-M. Chuang et al., Science **327**, 181 (2010)
8. A.V. Chubukov, D.M. Frenkel, Phys. Rev. B **46**, 11884 (1992)
9. A.V. Chubukov, D.V. Efremov, I. Eremin, Phys. Rev. B **78**, 134512 (2008)
10. A.V. Chubukov, Ann. Rev. Cond. Mat. Phys. **3**, 57–92 (2008)
11. R. Coldea, S.M. Hayden, G. Aeppli, T.G. Perring, C.D. Frost, T.E. Mason, S.-W. Cheong, Z. Fisk, Phys. Rev. Lett. **86**, 5377 (2001)
12. A.I. Coldea, J.D. Fletcher, A. Carrington, J.G. Analytis, A.F. Bangura, J.-H. Chu, A.S. Erickson, I.R. Fisher, N.E. Hussey, R.D. McDonald, Phys. Rev. Lett. 101, 216402 (2008); H. Shishido, A.F. Bangura, A.I. Coldea, S. Tonegawa, K. Hashimoto, S. Kasahara, P.M.C. Rourke, H. Ikeda, T. Terashima, R. Settai, Y. Onuki, D. Vignolles, C. Proust, B. Vignolle, A. McCollam, Y. Matsuda, T. Shibauchi, A. Carrington. Phys. Rev. Lett. **104**, 057008 (2010)
13. C. de la Cruz, Q. Huang, J.W. Lynn, J. Li, W. Ratcliff II, J.L. Zarestky, H.A. Mook, G.F. Chen, J.L. Luo, N.L. Wang, P. Dai. Nature **453**, 899 (2008)
14. V. Cvetkovic, Z. Tesanovic, Europhys. Lett. **85**, 37002 (2009)
15. M. Daghofer, Q. Luo, R. Yu, D. Yao, A. Moreo, E. Dagotto, Phys. Rev. B **81**, 180514(R) (2010)

16. S.O. Diallo, V.P. Antropov, T.G. Perring, C. Broholm, J.J. Pulikkotil, N. Ni, S.L. Budko, P. C. Canfield, A. Kreyssig, A.I. Goldman, R.J. McQueeney. Phys. Rev. Lett. **102**, 187206 (2009)
17. H. Ding, K. Nakayama, P. Richard, S. Souma, T. Sato, T. Takahashi, M. Neupane, Y.-M. Xu, Z.-H. Pan, A.V. Federov, Z. Wang, X. Dai, Z. Fang, G.F. Chen, J.L. Luo, N.L. Wang, J. Phys.: Condens. Matter **23**, 135701 (2011); P. Richard, T. Sato, K. Nakayama, T. Takahashi, H. Ding, Rep. Prog. Phys. **74**, 124512 (2011)
18. A. Dusza, A. Lucarelli, F. Pfuner, J.-H. Chu, I.R. Fisher, L. Degiorgi, EPL **93**, 37002 (2011)
19. I. Eremin, A.V. Chubukov, Phys. Rev. B **81**, 024511 (2010)
20. R.A. Ewings, T.G. Perring, J. Gillett, S.D. Das, S.E. Sebastian, A.E. Taylor, T. Guidi, A.T. Boothroyd, Phys. Rev. B **83**, 214519 (2011)
21. C. Fang, H. Yao, W.-F. Tsai, J.P. Hu, S.A. Kivelson, Phys. Rev. B **77**, 224509 (2008)
22. R.M. Fernandes, L.H. VanBebber, S. Bhattacharya, P. Chandra, V. Keppens, D. Mandrus, M.A. McGuire, B.C. Sales, A.S. Sefat, J. Schmalian, Phys. Rev. Lett. **105**, 157003 (2010)
23. R.M. Fernandes, A.V. Chubukov, J. Knolle, I. Eremin, J. Schmalian, Phys. Rev. B **85**, 024534 (2012)
24. R.M. Fernandes, A.V. Chubukov, J. Schmalian, Nature Phys. **10**, 97 (2014)
25. R.M. Fernandes, A.V. Chubukov (2016), arXiv:1607.00865v2
26. I.R. Fisher, L. Degiorgi, Z.X. Shen, Rep. Prog. Phys. **74**, 124506 (2011)
27. M.D. Frenkel, W. Hanke, Phys. Rev. B **42**, 6711 (1990)
28. S. Graser, T. A. Maier, P. J. Hirschfeld and D. J. Scalapino, New J. Phys. **11**, 025016 (2009)
29. P.J. Hirschfeld, M.M. Korshunov, I.I. Mazin, Rep. Prog. Phys. **74**, 124508 (2011)
30. Y. Kamihara, T. Watanabe, M. Hirano, H. Hosono, J. Am. Chem. Soc. **130**, 3296 (2008)
31. E. Kaneshita, T. Tohyama, Phys. Rev. B **82**, 094441 (2010)
32. E. Kaneshita, K. Tsutsui, T. Tohyama, Phys. Rev. B **84**, 020511(R) (2011)
33. L.V. Keldysh, YuV Kopaev, Sov. Phys. Solid State **6**, 2219 (1965)
34. A.F. Kemper, T.A. Maier, S. Graser, H.-P. Cheng, P.J. Hirschfeld, D.J. Scalapino, New J. Phys. 12, 073030 (2010); T.A. Maier, S. Graser, D.J. Scalapino, P.J. Hirschfeld. Phys. Rev. B **79**, 224510 (2009)
35. H.-H. Klauss, H. Luetkens, R. Klingeler, C. Hess, F.J. Litterst, M. Kraken, M.M. Korshunov, I. Eremin, S.-L. Drechsler, R. Khasanov, A. Amato, J. Hamann-Borreo, N. Leps, A. Kondrat, G. Behr, J. Werner, B. Büchner, Phys. Rev. Lett. **101**, 077005 (2008)
36. J. Knolle, I. Eremin, A.V. Chubukov, R. Moessner, Phys. Rev. B **81**, 140506 (2010)
37. J. Knolle, I. Eremin, A. Akbari, R. Moessner, Phys. Rev. Lett. **104**, 257001 (2010)
38. J. Knolle, I. Eremin, R. Moessner, Phys. Rev. B **83**, 224503 (2011)
39. M. Kovacic, M.H. Christensen, M.N. Gastiasoro, B.M. Andersen, Phys. Rev. B **91**, 064424 (2015)
40. F. Kruger, S. Kumar, J. Zaanen, J. van den Brink, Phys. Rev. B **79**, 054504 (2009)
41. K. Kubo, P. Thalmeier, J. Phys. Soc. Jpn. **78**, 083704 (2009)
42. Y. Li, Z. Yin, X. Wang, D.W. Tam, D. L. Abernathy, A. Podlesnyak, C. Zhang, M. Wang, L. Xing, C. Jin, K. Haule, G. Kotliar, T.A. Maier, P. Dai. Phys. Rev. Lett. **116**, 247001 (2016)
43. C. Liu, G.D. Samolyuk, Y. Lee, N. Ni, T. Kondo, A.F. Santander-Syro, S.L. Bud'ko, J.L. McChesney, E. Rotenberg, T. Valla, A.V. Fedorov, P.C. Canfield, B.N. Harmon, A. Kaminski, Phys. Rev. Lett. **101**, 177005 (2008)
44. D.H. Lu, M. Yi, S.-K. Mo, A.S. Erickson, J. Analytis, J.-H. Chu, D.J. Singh, Z. Hussain, T.H. Geballe, I.R. Fisher, Z.-X. Shen, Nature **455**, 81 (2008)
45. C. Liu, T. Kondo, R.M. Fernandes, A.D. Palczewski, E.D. Mun, N. Ni, A.N. Thaler, A. Bostwick, E. Rotenberg, J. Schmalian, S.L. Bud'ko, P.C. Canfield, A. Kaminski, Nature Physics **6**, 419 (2010)
46. W. Lv, F. Krüger, Ph Phillips, Phys. Rev. B **82**, 045125 (2010)
47. S. Maiti, M.M. Korshunov, T.A. Maier, P.J. Hirschfeld, A.V. Chubukov, Phys. Rev. B **84**, 224505 (2011)
48. S. Maiti, A.V. Chubukov, Phys. Rev. B **87**, 144511 (2013)
49. A. Moreo, M. Daghofer, J.A. Riera, E. Dagotto, Phys. Rev. B **79**, 134502 (2009)

50. S. Maiti, A.V. Chubukov, Superconductivity from repulsive interaction, in, *Proceedings of the XVII Training Course in the physics of Strongly Correlated Systems*. Vietri sul Mare (Salerno), Italy
51. F. Ahn, I. Eremin, J. Knolle, V.B. Zabolotnyy, S.V. Borisenko, B. Bchner, A.V. Chubukov, Phys. Rev. B **89**, 144513 (2014)
52. L. Boeri, O.V. Dolgov, A.A. Golubov, Phys. Rev. Lett. **101**, 026403 (2008)
53. I.I. Mazin, D.J. Singh, M.D. Johannes, M.H. Du, Phys. Rev. Lett. **101**, 057003 (2008)
54. T. Morinari, E. Kaneshita, T. Tohyama, Phys. Rev. Lett. **105**, 037203 (2010)
55. M. Nakajima, T. Liang, S. Ishida, Y. Tomioka, K. Kihou, C.H. Lee, A. Iyo, H. Eisaki, T. Kakeshita, T. Ito, S. Uchida, PNAS **108**, 12238 (2011)
56. C. Platt, C. Honerkamp, W. Hanke, New J. Phys. **11**, 055058 (2009)
57. Y. Qi, C. Xu, Phys. Rev. B **80**, 094402 (2009)
58. N.M.R. Peres, M.A.N. Arajo, Phys. Rev. B **65**, 132404 (2002)
59. S. Raghu, X.-L. Qi, C.-X. Liu, D.J. Scalapino, S.-C. Zhang. Phys. Rev. B **77**, 220503 (2008)
60. Y. Ran, F. Wang, H. Zhai, A. Vishwanath, Dung-Hai Lee, Phys. Rev. B **79**, 014505 (2009)
61. M.T. Rice, Phys. Rev. B **2**, 3619 (1970)
62. D.D. Scherer, I. Eremin, B.M. Andersen, Phys. Rev. B **94**, 180405 (2016)
63. J.R. Schrieffer, X.G. Wen, S.C. Zhang, Phys. Rev. B **39**, 11663 (1989)
64. Q. Si, E. Abrahams, Phys. Rev. Lett. **101**, 076401 (2008)
65. D.J. Singh, M.-H. Du, Phys. Rev. Lett. **100**, 237003 (2008)
66. A. Singh, Z. Tesanovic, Phys. Rev. B **41**, 614 (1990)
67. K. Sugimoto, E. Kaneshita, T. Tohyama, J. Phys. Soc. Jpn. **80**, 033706 (2011)
68. M.A. Tanatar, E.C. Blomberg, A. Kreyssig, M.G. Kim, N. Ni, A. Thaler, S.L. Budko, P.C. Canfield, A.I. Goldman, I.I. Mazin, R. Prozorov. Phys. Rev. B **81**, 184508 (2010)
69. K. Terashima, Y. Sekiba, J.H. Bowen, K. Nakayama, T. Kawahara, T. Sato, P. Richard, Y.-M. Xu, L.J. Li, G.H. Cao, Z.-A. Xu, H. Ding, T. Takahashi, Proc. Natl. Acad. Sci. **106**, 7330 (2009)
70. R. Thomale, Ch. Platt, W. Hanke, B.A. Bernevig, Phys. Rev. Lett. **106**, 187003 (2011)
71. Goetz S. Uhrig, Michael Holt, Jaan Oitmaa, Oleg P. Sushkov, R.R.P. Singh, Phys. Rev. B **79**, 092416 (2009)
72. B. Valenzuela, E. Bascones, M.J. Calderon, Phys. Rev. Lett. **105**, 207202 (2010)
73. F. Wang, H. Zhai, Y. Ran, A. Vishwanath, D.-H. Lee, Phys. Rev. Lett. **102**, 047005 (2009)
74. Y. Wang, A. Kreisel, V.B. Zabolotnyy, S.V. Borisenko, B. Büchner, T.A. Maier, P.J. Hirschfeld, D.J. Scalapino, Phys. Rev. B **88**, 174516 (2013)
75. X. Cenke, M. Muller, S. Sachdev, Phys. Rev. B **78**, 020501(R) (2008)
76. B.-J. Yang, H.-Y. Kee, Phys. Rev. B **82**, 195126 (2010)
77. L.X. Yang, Y. Zhang, H.W. Ou, J.F. Zhao, D.W. Shen, B. Zhou, J. Wei, F. Chen, M. Xu, C. He, Y. Chen, Z.D. Wang, X.F. Wang, T. Wu, G. Wu, X.H. Chen, M. Arita, K. Shimada, M. Taniguchi, Z.Y. Lu, T. Xiang, D.L. Feng, Phys. Rev. Lett. **102**, 107002 (2009)
78. M. Yi, D. Lu, J.-H Chu, J.G. Analytis, A.P. Sorini, A.F. Kemper, B. Moritz, S.-K. Mo, R.G. Moore, M. Hashimoto, W.-S. Lee, Z. Hussain, T.P. Devereaux, I.R. Fisher, Z.-X. Shen. PNAS **108**, 6878 (2011)
79. T. Yildirim, Phys. Rev. Lett. **101**, 057010 (2008)
80. Z.P. Yin, K. Haule, G. Kotliar, Nature Physics **10**, 845 (2014)
81. V.B. Zabolotnyy, D.S. Inosov, D.V. Evtushinsky, A. Koitzsch, A.A. Kordyuk, G.L. Sun, J.T. Park, D. Haug, V. Hinkov, A.V. Boris, C.T. Lin, M. Knupfer, A.N. Yaresko, B. Buechner, A. Varykhalov, R. Follath, S.V. Borisenko, Nature **457**, 569 (2009)
82. J. Zhao, D.T. Adroja, D.-X. Yao, R. Bewley, S. Li, X.F. Wang, G. Wu, X.H. Chen, J. Hu, P. Dai, Nature Phys. **5**, 555 (2009)

Chapter 3
Nematic Order and Fluctuations in Iron-Based Superconductors

Una Karahasanovic, Rafael M. Fernandes and Joerg Schmalian

Abstract The goal of these lecture notes is to provide a semi-phenomenological explanation of the phase diagram of iron based superconductors, and to summarize theoretical predictions for the signatures of the nematic degrees of freedom in various experimentally measurable quantities. In most cases we will focus on the spin-driven nematic approach. We will use this approach to

- Construct a phase diagram for the iron-based superconductors;
- Develop a φ^4 theory of the nematic phase and make the prediction about the nature of the nematic transition;
- Make predictions for the behaviour of the elastic modulus in the vicinity of a nematic/structural transition;
- Calculate the Raman response function, in the normal state and in the superconducting state;
- Calculate the resistivity anisotropy;
- Discuss the signatures of the Ising-nematic state in the magnetic spectrum.

Finally, we will briefly address the impact of nematic fluctuations on superconductivity.

3.1 Introduction

The analysis presented in these lecture notes closely follow work that we published together with our coauthors in [40–43, 53, 64, 65, 73].

U. Karahasanovic · J. Schmalian (✉)
Institut für Theorie der Kondensierten Materie and Institut für Festkörperphysik, Karlsruher Institut für Technologie, 76131 Karlsruhe, Germany
e-mail: joerg.schmalian@kit.edu

U. Karahasanovic
e-mail: una.karahasanovic@kit.edu

R.M. Fernandes
School of Physics and Astronomy, University of Minnesota, Minneapolis, MN 55455, USA
e-mail: rfernand@umn.edu

F. Mancini and R. Citro (eds.), *The Iron Pnictide Superconductors*,
Springer Series in Solid-State Sciences 186, DOI 10.1007/978-3-319-56117-2_3

3.1.1 In the Search for Nematic Phases

Different phases of matter can be classified according to which symmetries they break. Nematic phases – phases that break a rotational symmetry, but preserve translational symmetry – are known to occur in the context of liquid crystals [30]. Liquid crystals are made up of elongated molecules described as directors. In the nematic phase, these molecules are positioned randomly in space, preserving translational invariance, but they have a preferred direction, breaking the rotational symmetry.

In condensed matter physics, one is constantly confronted with new, exotic phases of matter. A very interesting question that arises is: Can similar phases, that break the rotational symmetry, but not the translational one, emerge in electronic systems? This question has generated a great deal of theoretical and experimental interest [28, 31, 32, 48–50, 55, 67, 92, 111, 112]. Signatures of the electron nematic phases have recently been observed in $2d$ electron gases in magnetic fields [48] and in the vicinity of striped phases of cuprates (for a review see for example [109]) and pnictides (for a review see [45]). They show anisotropies in transport measurements, can be detected in STM measurements, ARPES, neutron scattering experiments and NMR, and trigger simultaneous structural transitions and orbital ordering in pnictides.

The nematic instability can occur in (a) the charge (symmetric) channel or (b) the spin channel. Fermi surfaces of spin-nematic states display spin antisymmetric Pomeranchuk instabilities, which can be characterized by the angular momentum l. Spin nematics with even l break time reversal symmetry. We will not discuss these states in detail here, but rather concentrate on the nematic phases that occur in pnictides.

3.1.2 The Ising Nematic State in Iron-Based Superconductors: Brief Introduction, Current State of the Art and Open Questions

In what follows, we will focus on the so called Ising-nematic state in iron-based superconductors, which occurs above the striped spin-density wave phase and tracks the magnetic transition closely. The main feature of this phase is the lowering of the tetragonal symmetry to orthorhombic, while preserving spin-rotational invariance. Several theoretical proposals have been developed to explain its existence, mostly based on two different scenarios. One scenario proposes orbital ordering and fluctuations as a driving force behind the nematic transition [5, 74, 78, 82, 85, 86], while in the second scenario spin fluctuations associated with striped magnetic order generate the emergent electronic nematic order at temperatures above the Neel temperature [39, 42, 44, 45, 112]. We will take the latter approach. The primary argument for the spin-driven nematic approach is the vicinity or coincidence of the nematic and magnetic transition temperatures in the phase diagram. An interesting exception to this behavior is FeSe where there seems to be no magnetic order at ambient pressure. For a discussion of this material within a natural generalization of the theory that will be presented here, see [27].

The Ising-nematic degrees of freedom leave many experimental signatures, both in the tetragonal (disordered) and the nematic (ordered) phases: a large resistivity-anisotropy has been measured inside the nematic phase [23, 102]; a softening of the elastic modulus has been observed in a rather large temperature window in the tetragonal phase as one approaches the nematic/structural transition due to nematic fluctuations [11, 40, 43, 71, 72]. Strong signatures in the electronic Raman response in the normal, [51, 52, 58, 65, 68, 72, 73, 106, 115] and superconducting [9, 53, 58, 106] states (in the form of nematic resonance peak) have been observed. Moreover, anisotropies in other observables such as thermopower [63], optical conductivity [37, 90], torque magnetometry [66], STM measurements [96] and in NMR measurements [62] were observed.

Elastoresistance measurements [22] provide very strong evidence for electronic nematicity (rather than the nematic transition being a pure structural transition). Indeed, nematic degrees of freedom couple to the lattice [16, 40, 94] and as a consequence of this coupling induce the structural phase transition from the tetragonal into the orthorhombic phase. Strong evidence for the spin-driven nematic scenario comes from the scaling of the shear modulus and the NMR spin-lattice relaxation rate for 122 systems, discussed in [43]. We also mention that the nematic order was observed in FeSe, a material where spin-magnetic order is absent at ambient pressure (and can only be generated via application of external pressure), while a structural transition along with a softening of the lattice occurs at ambient pressure in a manner very similar to the 122 iron pnictides [6, 12, 105]. This observation led to the proposal that orbital fluctuations might play an important role in FeSe and drive the nematic transition. However, an alternative proposal, based on a straightforward extension of the spin-nematic scenario, in which imaginary charge density wave fluctuations drive the transition has also been proposed [27]. What makes FeSe different from the, for example 122 iron pnictides is its very small Fermi surface, as observed in the experiments of [104, 110]. In this limit, almost degenerate imaginary charge density and spin density waves are expected [25–27, 88], and imaginary charge density wave fluctuations (in cooperation with magnetic fluctuations) therefore seem a plausible candidate to drive the nematic transition. Clearly further experimental investigation is needed to judge weather this or other scenarios apply to FeSe. Another important distinguishing feature of FeSe is that, according to first principle calculations, it has a rather extended magnetic degeneracy with competing ground states with ordering vectors (π, q) and (q, π), with $q < \pi/2$. Such an enhanced degeneracy was shown to not only increase the nematic transition temperature, but also to suppress the magnetic transition temperature [54].

3.2 Summary of Collective Field Theories of Magnetism

The theory presented in the subsequent Section is largely based on weak coupling expansions of the partition function, leading to collective degrees of freedom. These collective field theories are expected to be valid even in the strong coupling limit. Nevertheless, the ability to explicitly derive the terms of the Ginzburg–Landau expansions within the weak-coupling approach makes it a very suitable formulation for a summer school.

3.2.1 *Hertz-Millis Theory of Magnetism*

In order to obtain a first impression of the description of collective magnetism in itinerant systems we summarize the order-parameter approach to the problem put forward by Hertz and Millis [56, 89]. We start from the following Hamiltonian of interacting fermions

$$\begin{aligned} H &= \int_{\mathbf{k}} \varepsilon_{\mathbf{k}} c^{\dagger}_{\mathbf{k}\alpha} c_{\mathbf{k}\alpha} - \frac{J}{2} \int_{\mathbf{k},\mathbf{k}',\mathbf{q}} c^{\dagger}_{\mathbf{k}\alpha} \boldsymbol{\sigma}_{\alpha\beta} c_{\mathbf{k}+\mathbf{q}\beta} \cdot c^{\dagger}_{\mathbf{k}'\gamma} \boldsymbol{\sigma}_{\gamma\delta} c_{\mathbf{k}'-\mathbf{q}\delta} \\ &= \int_{\mathbf{k}} \varepsilon_{\mathbf{k}} c^{\dagger}_{\mathbf{k}\alpha} c_{\mathbf{k}\alpha} - \frac{J}{2} \int_{\mathbf{q}} \mathbf{s}_{\mathbf{q}} \cdot \mathbf{s}_{-\mathbf{q}}, \end{aligned} \tag{3.1}$$

where $\mathbf{s}_{\mathbf{q}} = \int_{\mathbf{k}} c^{\dagger}_{\mathbf{k}\alpha} \sigma_{\alpha\beta} c_{\mathbf{k}+\mathbf{q}\beta}$ is the fermionic spin density. Here $c_{\mathbf{k}\alpha}$ is the annihilation operator of a fermion in a single particle state with momentum $\mathbf{k}$ and spin α. $c^{\dagger}_{\mathbf{k}\alpha}$ is the corresponding creation operator. $\varepsilon_{\mathbf{k}}$ is the band structure of the fermions that we assume to form only one band crossing the Fermi energy. J is a magnetic interaction, while $\boldsymbol{\sigma} = (\sigma^x, \sigma^y, \sigma^z)$ stands for the vector of the Pauli matrices. We also used the short hand notation $\int_{\mathbf{q}} \cdots = \int_{BZ} \frac{d^d q}{(2\pi)^d} \cdots$ for $d-$dimensional momentum integrals that go over the first Brillouin zone.

To proceed we use the functional integral representation of this Hamiltonian in terms of Grassmann variables (see [4, 91]) which leads to the partition function

$$Z = \int Dc^{\dagger} Dc e^{-S}, \tag{3.2}$$

with action

$$S = \int_0^{\beta} d\tau \left(- \int_{\mathbf{k}} c^{\dagger}_{\mathbf{k}\alpha}(\tau)\, \partial_{\tau} c_{\mathbf{k}\alpha}(\tau) + H\left[c^{\dagger}_{\mathbf{k}\alpha}(\tau), c_{\mathbf{k}\alpha}(\tau)\right] \right). \tag{3.3}$$

Here τ is the imaginary time that varies from $\tau = 0$ to $\tau = \beta$, where $\beta = 1/(k_B T)$ is the inverse temperature (we set $k_B = 1$ in the following). To simplify the notation, we do not introduce new symbols for the Grassmann variables (which obey the antiperiodic boundary conditions $c_{\mathbf{k}\alpha}^{(\dagger)}(\tau+\beta)=-c_{\mathbf{k}\alpha}^{(\dagger)}(\tau)$) and the corresponding annihilation and creation operators.

For our subsequent analysis we only need to know one result for Grassmann integrals [91]:

$$\int Dc^{\dagger}Dce^{\sum_{ij}c_i^{\dagger}\mathcal{G}_{ij}^{-1}c_j+\sum_j\left(\eta_j c_j^{\dagger}+\eta_j^{\dagger}c_j\right)}=e^{\mathrm{tr}\log(-\mathcal{G}^{-1})-\sum_{ij}\eta_i^{\dagger}\mathcal{G}_{ij}\eta_j}, \tag{3.4}$$

where i and j stand for all possible arguments of the Grassman variables c_i, η_i, etc. Thus, if the exponent of a fermionic integral contains terms up to second order, one can always explicitly evaluate the integral.

It is useful to perform the Fourier transform:

$$\begin{aligned} c_{\mathbf{k}\alpha}(\omega_n)&=\int_0^{\beta}d\tau e^{i\omega_n\tau}c_{\mathbf{k}\alpha}(\tau),\\ c_{\mathbf{k}\alpha}(\tau)&=T\sum_{n=-\infty}^{\infty}e^{-i\omega_n\tau}c_{\mathbf{k}\alpha}(\omega_n), \end{aligned} \tag{3.5}$$

where $\omega_n=(2n+1)\pi T$ is an odd multiple of πT, which guarantees the antiperiodic boundary conditions (n is an integer). The ω_n are fermionic Matsubara frequencies. Analogously we use

$$\begin{aligned} c_{\mathbf{k}\alpha}^{\dagger}(\omega_n)&=\int_0^{\beta}d\tau e^{-i\omega_n\tau}c_{\mathbf{k}\alpha}^{\dagger}(\tau),\\ c_{\mathbf{k}\alpha}^{\dagger}(\tau)&=T\sum_{n=-\infty}^{\infty}e^{i\omega_n\tau}c_{\mathbf{k}\alpha}^{\dagger}(\omega_n) \end{aligned} \tag{3.6}$$

for the Grassmann variables that represent creation operator. Inserting these Fourier expansions yields for example

$$\int_0^{\beta}d\tau\int_{\mathbf{k}}c_{\mathbf{k}\alpha}^{\dagger}(\tau)\,\partial_{\tau}c_{\mathbf{k}\alpha}(\tau)=T\sum_n\int_{\mathbf{k}}c_{\mathbf{k}\alpha}^{\dagger}(\omega_n)\,i\omega_n c_{\mathbf{k}\alpha}(\omega_n), \tag{3.7}$$

or

$$\int_0^{\beta}\int_{\mathbf{q}}\mathbf{s}_{\mathbf{q}}(\tau)\cdot\mathbf{s}_{-\mathbf{q}}(\tau)=T\sum_m\int_{\mathbf{q}}\mathbf{s}_{\mathbf{q}}(\nu_m)\cdot\mathbf{s}_{-\mathbf{q}}(-\nu_m), \tag{3.8}$$

where

$$\mathbf{s}_{\mathbf{q}}(\nu_m)=T\sum_n\int_{\mathbf{k}}c_{\mathbf{k}\alpha}^{\dagger}(\omega_n)\,\sigma_{\alpha\beta}c_{\mathbf{k}+\mathbf{q}\beta}(\omega_n+\nu_m).$$

To ensure that all arguments of the fermionic Grassmann variables are odd multiples of πT, it must hold that $\nu_m=2m\pi T$, with integer m. The ν_m are bosonic Matsubara frequencies, reflecting the fact that the behavior of fermionic bilinears is similar to bosonic degrees of freedom as the boundary conditions $\mathbf{s_q}(\tau+\beta)=\mathbf{s_q}(\tau)$ are periodic.

Because of the symmetric appearance of momentum and Matsubara frequencies, we simplify our notation defining $q=(\mathbf{q},\nu_m)$ or $k=(\mathbf{k},\omega_n)$ and $\int_q\cdots=T\sum_n\int\frac{d^dq}{(2\pi)^d}\cdots$ and analogous for $\int_k\cdots$. Thus, we write:

$$T\sum_m\int_{\mathbf{q}}\mathbf{s_q}(\nu_m)\cdot\mathbf{s_{-q}}(-\nu_m)=\int_q\mathbf{s}_q\cdot\mathbf{s}_{-q}. \tag{3.9}$$

This yields for the action

$$S=-\int_k c^\dagger_{k\alpha}(i\omega_n-\varepsilon_\mathbf{k})\,c_{k\alpha}-\frac{J}{2}\int_q\mathbf{s}_q\cdot\mathbf{s}_{-q}. \tag{3.10}$$

Next we perform an exact Hubbard–Stratonovich transformation that concerns the interaction part of the action and rewrite:

$$e^{\frac{J}{2}\int_q\mathbf{s}_q\cdot\mathbf{s}_{-q}}\propto\int D\phi e^{-\frac{J}{2}\int_q\boldsymbol{\phi}_q\cdot\boldsymbol{\phi}_{-q}-J\int_q\boldsymbol{\phi}_q\cdot\mathbf{s}_{-q}}, \tag{3.11}$$

where $\boldsymbol{\phi}_q=\boldsymbol{\phi}_\mathbf{q}(\nu_n)$ is a three-component scalar field with bosonic boundary conditions. Thus, we obtain an effective action of fermions. Now, fermions no longer interact directly, but instead with the collective bosonic degree of freedom $\boldsymbol{\phi}_q$:

$$S_{\mathrm{eff}}=-\int_k(i\omega_n-\varepsilon_\mathbf{k})\,c^\dagger_{k\alpha}c_{k\alpha}+J\int_{k,k'}c^\dagger_{k\alpha}\sigma_{\alpha\beta}c_{k'\beta}\cdot\phi_{k-k'}+\frac{J}{2}\int_q\phi_q\cdot\phi_{-q}. \tag{3.12}$$

The appeal of this formulation is that techniques that were developed for fermions interacting with bosons (e.g. photons or phonons) can be applied to this model. It is also a natural step to develop a theory for interacting collective order parameter degrees of freedom. A potential drawback of this formulation is however that it is a rather biased approach. Emphasis is given to magnetic ordering or magnetic fluctuations and all other degrees of freedom are considered secondary. This can, but does not have to be the right point of view.

In view of our goal to develop an order parameter theory we can, at least formally, integrate out the fermions and obtain an effective theory that consists only of bosonic degrees of freedom. This is formally possible as we can express the fermionic part in a quadratic form:

$$S_{\mathrm{col}}=-\int_{kk'}c^\dagger_{k\alpha}\mathcal{G}^{-1}_{k\alpha,k'\beta}c_{k'\beta}+\frac{J}{2}\int_q\phi_q\cdot\phi_{-q}, \tag{3.13}$$

with

$$\mathscr{G}^{-1}_{k\alpha,k'\beta} = (i\omega_n - \varepsilon_{\mathbf{k}})\,\delta_{\alpha\beta}\delta_{k,k'} - J\phi_{k-k'}\sigma_{\alpha\beta}, \tag{3.14}$$

which plays the role of the inverse fermion Green's function in an external magnetic field that fluctuates in time and space. With this quadratic form we can use (3.4) and perform the integration over the fermionic variables. In this approach one must be careful as one integrates out massless fermions that may result in singularities of the resulting bosonic theory. Let us first simply evaluate the Gaussian integration over the fermionic degrees of freedom and discuss the potential danger of this approach later. Using (3.4), we obtain for the action

$$S[\phi] = \frac{J}{2}\int_q \phi_q \cdot \phi_{-q} - \operatorname{tr}\log\left(-\mathscr{G}^{-1}\right). \tag{3.15}$$

We split the fermionic Green's function as

$$\mathscr{G}^{-1} = \mathscr{G}_0^{-1} - \mathscr{V}, \tag{3.16}$$

where $\mathscr{G}^{-1}_{0,k\alpha,k'\beta} = G_k^{-1}\delta_{\alpha\beta}\delta_{k,k'}$ and $\mathscr{V}_{k\alpha,k'\beta} = J\phi_{k-k'}\sigma_{\alpha\beta}$. Here

$$G_k = \frac{1}{i\omega_n - \varepsilon_{\mathbf{k}}} \tag{3.17}$$

is the fermion propagator of a free particle with dispersion $\varepsilon_{\mathbf{k}}$. This allows to express the second term in the bosonic action $S[\phi]$ as

$$\begin{aligned}
\operatorname{tr}\log\left(-\mathscr{G}^{-1}\right) &= \operatorname{tr}\log\left(-\mathscr{G}_0^{-1}(1-\mathscr{G}_0\mathscr{V})\right) \\
&= \operatorname{tr}\log\left(-\mathscr{G}_0^{-1}\right) + \operatorname{tr}\log\left(1-\mathscr{G}_0\mathscr{V}\right) \\
&= \beta F_0 - \sum_{n=1}^{\infty}\frac{1}{n}\operatorname{tr}\left(\mathscr{G}_0\mathscr{V}\right)^n,
\end{aligned} \tag{3.18}$$

where $F_0 = \operatorname{tr}\log\left(-\mathscr{G}_0^{-1}\right)$ is the free energy of an ideal Fermi gas with dispersion $\varepsilon_{\mathbf{k}}$. Since $\mathscr{G}_0$ is diagonal in spin space and $\mathscr{V} \propto \boldsymbol{\sigma}$, it follows that the trace in spin space vanishes for $n = 1$. The first nontrivial contribution occurs for $n = 2$. It holds

$$\begin{aligned}
\operatorname{tr}\left(\mathscr{G}_0\mathscr{V}\right)^2 &= J^2\int_{k,k'} G_k\phi^i_{k-k'}G_{k'}\,\phi^j_{k'-k}\operatorname{tr}\left(\sigma^i\sigma^j\right) \\
&= -\int_q \Pi(q)\,\phi_q\cdot\phi_{-q},
\end{aligned} \tag{3.19}$$

where we used $\operatorname{tr}\left(\sigma^i\sigma^j\right) = 2\delta_{ij}$ and introduced:

$$\Pi(q) = -2J^2\int_k G_k G_{k-q}. \tag{3.20}$$

For $n = 3$ we have

$$\mathrm{tr}\,(\mathscr{G}_0\mathscr{V})^3 = J^3 \int_{k_1,k_2,k_3} G_{k_1}\phi^i_{k_1-k_2} G_{k_2}\phi^j_{k_2-k_3} G_{k_3}\phi^l_{k_3-k_1} \mathrm{tr}\left(\sigma^i\sigma^j\sigma^l\right). \tag{3.21}$$

It holds $\mathrm{tr}\left(\sigma^i\sigma^j\sigma^l\right) = \delta_{ij}\mathrm{tr}\sigma^l + i\varepsilon_{ijm}\mathrm{tr}\left(\sigma^m\sigma^l\right) = 2i\varepsilon_{ijl}$ and it follows

$$\begin{aligned}\mathrm{tr}\,(\mathscr{G}_0\mathscr{V})^3 &= 2iJ^3 \int_{k,q_1,q_2} G_k G_{k-q_1} G_{k-q_1-q_2}\left(\phi_{q_1}\times\phi_{q_2}\right)\cdot\phi_{-q_1-q_2} \\ &= i\int_k \gamma\,(q_i)\left(\phi_{q_1}\times\phi_{q_2}\right)\cdot\phi_{-q_1-q_2},\end{aligned} \tag{3.22}$$

where

$$\gamma\,(q_i) = 2J^3\int_k G_k G_{k-q_1} G_{k-q_1-q_2}. \tag{3.23}$$

Finally, we analyze the term with $n = 4$:

$$\begin{aligned}\mathrm{tr}\,(\mathscr{G}_0\mathscr{V})^4 &= J^4 \int_{k_1\cdots k_4} G_{k_1}\phi^i_{k_1-k_2} G_{k_2}\phi^j_{k_2-k_3} G_{k_3}\phi^l_{k_3-k_4} G_{k_4}\phi^m_{k_4-k_1}\mathrm{tr}\left(\sigma^i\sigma^j\sigma^l\sigma^m\right) \\ &= 2J^4\int_{k,q_1\cdots q_3} G_k G_{k-q_1} G_{k-q_1-q_2} G_{k-q_1-q_2-q_3} \\ &\quad\times\left(\phi_{q_1}\cdot\phi_{q_2}\phi_{q_3}\cdot\phi_{q_4} + \phi_{q_2}\cdot\phi_{q_3}\phi_{q_1}\cdot\phi_{q_4} - \phi_{q_1}\cdot\phi_{q_3}\phi_{q_2}\cdot\phi_{q_4}\right),\end{aligned} \tag{3.24}$$

where $q_4 = -q_1 - q_2 - q_3$ and we used:

$$\begin{aligned}\mathrm{tr}\left(\sigma^i\sigma^j\sigma^l\sigma^m\right) &= \delta_{ij}\mathrm{tr}\left(\sigma^l\sigma^m\right) + i\varepsilon_{ijn}\mathrm{tr}\left(\sigma^n\sigma^l\sigma^m\right) \\ &= 2\delta_{ij}\delta_{lm} - 2\varepsilon_{ijn}\varepsilon_{nlm} \\ &= 2\left(\delta_{ij}\delta_{lm} + \delta_{im}\delta_{jl} - \delta_{il}\delta_{jm}\right).\end{aligned} \tag{3.25}$$

The last result can we written as

$$\mathrm{tr}\,(\mathscr{G}_0\mathscr{V})^4 = \int_{k,q_1\cdots q_3} u\,(q_i)\,\phi_{q_1}\cdot\phi_{q_2}\phi_{q_3}\cdot\phi_{q_4}, \tag{3.26}$$

where

$$u\,(q_i) = 2J^4\int_k\left[\left(G_{k-q_1} + G_{k-q_2}\right)G_{k-q_1-q_2} - G_{k-q_1}G_{k-q_1-q_3}\right]G_k G_{k-q_1-q_2-q_3}. \tag{3.27}$$

It is now possible to write down an effective action for the bosonic degrees of freedom

$$S = \frac{1}{2}\int_q (J - \Pi(q))\,\phi_q \cdot \phi_{-q} + \frac{i}{3}\int_{q_1 q_2} \gamma(q_i)\left(\phi_{q_1} \times \phi_{q_2}\right) \cdot \phi_{-q_1-q_2}$$
$$+\frac{1}{4}\int_{q_1 \cdots q_3} u(q_i)\,\phi_{q_1} \cdot \phi_{q_2}\phi_{q_3} \cdot \phi_{-q_1-q_2-q_3}. \tag{3.28}$$

The term with $n = 2$ corresponds to a renormalization of the bare boson propagator, the other terms correspond to interactions between collective spin modes. The fact that we were courageous enough to integrate out massless degrees of freedom (fermions) leads to singular behavior of $\Pi(q)$, $\gamma(q)$, etc. To get a better understanding of such singularities, let us evaluate the individual terms $\Pi(q)$, $\gamma(q_i)$ and $u(q_i)$. We start with $\Pi(q)$ that is explicitly given as:

$$\Pi_{\mathbf{q}}(i\omega_n) = -2J^2 T \sum_m \int \frac{d^d k}{(2\pi)^d} \frac{1}{i\Omega_m - \varepsilon_{\mathbf{k}}} \frac{1}{i(\Omega_m + \omega_n) - \varepsilon_{\mathbf{k+q}}}. \tag{3.29}$$

For momenta $q \ll k_F$ we can expand $\varepsilon_{\mathbf{k+q}} = \varepsilon_{\mathbf{k}} + \mathbf{v_k} \cdot \mathbf{q}$. Assuming $\mathbf{v_k} = v\frac{\mathbf{k}}{|k|}$, as holds for a spherical Fermi surface one obtains:

$$\Pi(q, i\Omega) = 2J^2 \rho_F \left(1 - \frac{|\Omega|}{\sqrt{(vq)^2 + \Omega^2}}\right). \tag{3.30}$$

We now see that this particle hole diagram depends on the order of limits

$$\Pi(q = 0, i\Omega \to 0) = 0,$$
$$\Pi(q \to 0, i\Omega = 0) = 2J\rho_F. \tag{3.31}$$

The first limit reflects that there is an underlying conservation law (spin conservation). The analytic continuation to the real frequency axis $i\Omega \to \omega + i0^+$ yields the retarded function

$$\Pi^R(q, \omega) = 2J^2 \rho_F \left(1 - \frac{|\omega|}{\sqrt{\omega^2 - (vq)^2}}\theta(|\omega| - vq) + i\frac{|\omega|}{\sqrt{(vq)^2 - \omega^2}}\theta(vq - |\omega|)\right). \tag{3.32}$$

The imaginary part of Π^R is finite for arbitrarily small frequencies, reflecting the fact that spin excitations can decay in particle-hole excitations. This phenomenon is called Landau damping.

The nature of the Landau damping is different in case of an antiferromagnet. To this end we analyze $\Pi_{\mathbf{q}}(i\omega_n)$ for large momentum $\mathbf{q} \approx \mathbf{Q}$ with antiferromagnetic ordering vector $\mathbf{Q}$. Let us determine $\Pi(\mathbf{Q}, \Omega)$.

We expect a large contribution to Landau damping from those momentum states on the Fermi surface that can be connected to other Fermi surface states by the magnetic vector $\mathbf{Q}$. Implicitly, of course we assume that such momenta exist, which are called hot spots on the Fermi surface. Near such hot spots we expand:

$$\varepsilon_{\mathbf{k}} \approx \varepsilon_{\mathbf{k}_h} + \mathbf{v}_{\mathbf{k}_h} \cdot (\mathbf{k} - \mathbf{k}_h) ,$$
$$\varepsilon_{\mathbf{k}+\mathbf{Q}} \approx \varepsilon_{\mathbf{k}_h+\mathbf{Q}} + \mathbf{v}_{\mathbf{k}_h+\mathbf{Q}} \cdot (\mathbf{k} - \mathbf{k}_h) . \tag{3.33}$$

Since $\mathbf{k}_h$ and $\mathbf{k}_h + \mathbf{Q}$ are, by assumption, states on the Fermi surface, it follows $\varepsilon_{\mathbf{k}_h} = \varepsilon_{\mathbf{k}_h+\mathbf{Q}} = 0$. Let us analyze a two-dimensional Fermi surface and assume that the two velocities are not collinear. In a coordinate system where the y–axis points along $\mathbf{Q}$ holds with $\mathbf{p} = \mathbf{k} - \mathbf{k}_h$:

$$\mathbf{v}_{\mathbf{k}_h} = v \left(\cos \frac{\varphi}{2} p_x, \sin \frac{\varphi}{2} p_y \right) ,$$
$$\mathbf{v}_{\mathbf{k}_h+\mathbf{Q}} = v \left(\cos \frac{\varphi}{2} p_x, - \sin \frac{\varphi}{2} p_y \right) ,$$

where φ is the angle between the two velocities. Transforming to

$$p_1 = \cos \frac{\varphi}{2} p_x - \sin \frac{\varphi}{2} p_y,$$
$$p_2 = \cos \frac{\varphi}{2} p_x + \sin \frac{\varphi}{2} p_y.$$

It follows that at $T = 0$:

$$\Pi_{\mathbf{Q}} (i\omega) = - \frac{2J^2}{|\sin \varphi|} \int \frac{dp_1 dp_2 d\Omega}{(2\pi)^3} \frac{1}{i\Omega - v p_1} \frac{1}{i (\Omega + \omega) - v p_2} . \tag{3.34}$$

We use

$$\int dp \frac{1}{i\Omega - vp} = \frac{-i\pi \operatorname{sign} (\Omega)}{v} , \tag{3.35}$$

to obtain

$$\begin{aligned} \delta \Pi_{\mathbf{Q}} (i\omega) &= \Pi_{\mathbf{Q}} (i\omega) - \Pi_{\mathbf{Q}} (0) \\ &= \frac{J^2}{2v^2 |\sin \varphi|} \int_{-\infty}^{\infty} \frac{d\Omega}{2\pi} \operatorname{sign} (\Omega) (\operatorname{sign} (\omega + \Omega) - \operatorname{sign} (\omega)) \\ &= \frac{J^2}{4\pi v^2 |\sin \varphi|} |\omega| . \end{aligned} \tag{3.36}$$

This can also be analytically continued to the real frequency axis and yields for the imaginary part

$$\operatorname{Im} \Pi^R (\omega) = - \frac{J^2}{4\pi v^2 |\sin \varphi|} \omega . \tag{3.37}$$

Once again, a spin excitation can decay into particle-hole excitations at arbitrarily small energy. The difference between ferromagnetic and antiferromagnetic fluctuations manifests itself in the momentum dependence of the Landau damping.

We can also briefly analyze the higher order terms. For an arbitrary density of states holds:

$$\gamma(0) = 2g^3 T \sum_m \int d\varepsilon \frac{\rho(\varepsilon)}{(i\omega_m - \varepsilon)^3} = 2g^3 \int \frac{d\varepsilon}{8T^2} \rho(\varepsilon) \frac{\tanh\left(\frac{\varepsilon}{2T}\right)}{\cosh^2\left(\frac{\varepsilon}{2T}\right)}. \tag{3.38}$$

For $T \to 0$ follows that

$$\gamma(0) = g^3 \rho'_F, \tag{3.39}$$

where ρ'_F is the first derivative of the density of states evaluated at the Fermi energy. Thus, contributions from low energies, where the density of states is constant, do not affect γ. Of course, a constant contribution to the three point interaction is irrelevant and the Fourier transform is

$$\gamma(0) \int_x (\phi_x \times \phi_x) \cdot \phi_x = 0. \tag{3.40}$$

In some situations, it is however important to include the non-local behaviour of the cubic term, see [99].

It is now possible to write down an effective action for the bosonic degrees of freedom

$$S = \frac{1}{2} \int_q (g - \Pi(q)) \phi_q \cdot \phi_{-q} + \frac{i}{3} \int_{q_1 q_2} \gamma(q_i) \left(\phi_{q_1} \times \phi_{q_2}\right) \cdot \phi_{-q_1-q_2}$$
$$+ \frac{1}{4} \int_{q_1 \cdots q_3} u(q_i) \phi_{q_1} \cdot \phi_{q_2} \phi_{q_3} \cdot \phi_{-q_1-q_2-q_3}. \tag{3.41}$$

Thus, one can microscopically generate an order-parameter theory of a quantum many body theory.

3.2.2 *Introducing the Large-N Approach*

In the previous section we have seen that one can motivate a Ginzburg–Landau type expansion from a microscopic Hamiltonian. The next task is to solve the theory in some sensible limit. In view of our subsequent analysis, we use the large-N expansion, where N is the number of components of the vector ϕ.

We start from the action $S = S_{\text{static}} + S_{\text{dyn}}$ with

$$S_{\text{static}}[\phi] = \frac{1}{2} \int d^d x \int d\tau \left[r_0 \phi^2(\mathbf{x}, \tau) + (\nabla \phi(\mathbf{x}, \tau))^2 + \frac{u}{2N} \phi^4(\mathbf{x}, \tau) \right], \tag{3.42}$$

and

$$S_{\text{dyn}} = \frac{\gamma_z T}{2} \sum_n \int d^d x |\omega_n|^{2/z} \phi(\mathbf{x}, i\omega_n) \cdot \phi(\mathbf{x}, -i\omega_n). \tag{3.43}$$

$\phi = (\phi_1, \ldots, \phi_N)$ is an N-component vector. For $z = 1$ this corresponds to a ballistic spin dynamics with $\gamma_1 = c^{-2}$ given by some characteristic velocity $c \simeq Ja_0$ with exchange constant J and typical interparticle distance a_0. This is relevant for insulating antiferromagnets or antiferromagnets without hot-spots. In the case of $z = 2$, the spin dynamics is overdamped (as in the case of antiferromagnetic fluctuations with hot spots) and $\gamma_2 = \Gamma^{-1}$ with $\Gamma \simeq E_F a_0^2$ and Fermi energy E_F.

The solution of the problem in the large N limit can be obtained via Hubbard–Stratonovich transformation:

$$S[\phi, \lambda] = S_{\text{dyn}} - \int d^d x \int d\tau \frac{N}{4u} \lambda^2(\mathbf{x}, \tau) \tag{3.44}$$

$$+\frac{1}{2} \int d^d x \int d\tau \left[(\nabla\phi(\mathbf{x}, \tau))^2 + (r_0 + \lambda(\mathbf{x}, \tau)) \phi^2(\mathbf{x}, \tau) \right]. \tag{3.45}$$

Integrating out ϕ, leads to $Z = \int D\lambda \exp(-\mathscr{S}[\lambda])$ with

$$\mathscr{S}[\lambda] = N \left(\frac{1}{2} \text{tr} \log\left(\mathscr{D}^{-1}\right) - \frac{1}{4u} \int d^D x \int d\tau \lambda^2(\mathbf{x}, \tau) \right), \tag{3.46}$$

where

$$\mathscr{D}^{-1}\left(\mathbf{x}, \mathbf{x}', \tau - \tau'\right) = \alpha\left(\tau - \tau'\right) \delta\left(\mathbf{x} - \mathbf{x}'\right) + \left(r_0 - \nabla_{\mathbf{x}}^2 - \lambda(\mathbf{x})\right) \delta\left(\tau - \tau'\right) \delta\left(\mathbf{x} - \mathbf{x}'\right). \tag{3.47}$$

$\alpha(\tau)$ is the Fourier transform of $\gamma_z |\omega_n|^{2/z}$, i.e. $\alpha\left(\tau - \tau'\right) \sim \Gamma^{-1}\left(\tau - \tau'\right)^{-2}$ for $z = 2$ and $\alpha\left(\tau - \tau'\right) \sim \frac{\partial^2}{c^2 \partial \tau^2} \delta\left(\tau - \tau'\right)$ for $z = 1$. For large N, the λ integration is performed via saddle point approximation and we find

$$\lambda(\mathbf{x}) = u\mathscr{D}\left(\mathbf{x}, \mathbf{x}, 0^+\right) = u\left\langle \phi^2(\mathbf{x}) \right\rangle. \tag{3.48}$$

If one is interested in the symmetry broken phase one can separate the components of ϕ into one component σ and $N - 1$ components π and integrate only over π. This gives after adding an external field along the σ-direction

$$\begin{aligned} S[\sigma, \lambda] = {} & S_{\text{dyn}}[\sigma] - h \int d^d x \int d\tau \sigma(\mathbf{x}, \tau) \\ & - \int d^d x \int d\tau \frac{N}{4u} \lambda^2(\mathbf{x}, \tau) + \frac{N-1}{2} \text{tr} \log\left(\mathscr{D}^{-1}\right) \\ & + \frac{1}{2} \int d^d x \int d\tau \left[(\nabla\sigma(\mathbf{x}, \tau))^2 + (r_0 + \lambda(\mathbf{x}, \tau)) \sigma^2(\mathbf{x}, \tau) \right]. \end{aligned} \tag{3.49}$$

A static saddle point is now given by

$$h = -\nabla^2 \sigma(\mathbf{x}) + (r_0 + \lambda(\mathbf{x}))\sigma(\mathbf{x}),$$
$$\lambda(\mathbf{x}) = u\left(\mathscr{D}\left(\mathbf{x}, \mathbf{x}, 0^+\right) + \sigma^2(\mathbf{x})/N\right). \tag{3.50}$$

Assuming a translation invariant solution, it holds (at $T = 0$)

$$\lambda = u \int \frac{d^d q d\omega}{(2\pi)^{D+1}} \frac{1}{q^2 + \gamma_z |\omega|^{2/z} + r_0 + \lambda} + \sigma^2/N,$$
$$(r_0 + \lambda)\sigma = h. \tag{3.51}$$

The correlation function is given as $D(k,\omega)^{-1} = q^2 + \gamma_z |\omega|^{2/z} + \xi^{-2}$, such that $r = \xi^{-2} = r_0 + \lambda$ and r and σ are determined by:

$$r = r_0 + u \int_p \frac{1}{p^2 + r} + \sigma^2/N$$
$$r\sigma = h, \tag{3.52}$$

where $p^2 \equiv q^2 + |\omega|^{2/z}$ and $\int_p = \int \frac{d^{d+z}q}{(2\pi)^{d+z}}$. For simplicity, we use frequency units with $\gamma_z = 1$ and set the conjugate field to zero. A quantum phase transition happens for $r \to 0$, i.e. when the coupling constant $\bar{u} = -u/r_0$ obeys

$$1 = \bar{u}_c \int_p \frac{1}{p^2}, \tag{3.53}$$

which converges for $d + z > 2$. This gives

$$\frac{r}{|r_0|} = (\bar{u} - \bar{u}_c) \int_p \frac{1}{p^2 + \xi^{-2}} - \bar{u}_c r \int_p \frac{1}{(p^2 + r)\, p^2}. \tag{3.54}$$

If $d + z > 4$, the last integral is convergent and the first term dominates. It follows

$$\xi \propto (\bar{u} - \bar{u}_c)^{-1/2} \tag{3.55}$$

If $2 < d + z < 4$ it holds $\int_p \frac{r}{(p^2+r)p^2} \propto r^{\frac{d+z-2}{2}}$, which dominates over the l.h.s., yielding

$$\xi \propto (\bar{u} - \bar{u}_c)^{-1/(d+z-2)}. \tag{3.56}$$

The static susceptibility is given by

$$\chi(q = 0, \omega = 0) = \xi^2 \tag{3.57}$$

which gives $\gamma = 2\nu$.

In the ordered phase, $\sigma \neq 0$, it holds $\xi^{-2} = 0$ due to Goldstone theorem. It follows

$$\sigma^2 = N\left(|r_0| - u\int_p \frac{1}{p^2}\right) \tag{3.58}$$

which gives

$$\sigma = \sigma_0\left(\bar{u}_c - \bar{u}\right)^{1/2} \tag{3.59}$$

with $\sigma_0 = \sqrt{N\frac{|r_0|}{\bar{u}_c}}$.

These results demonstrate that the large-N analysis is an efficient, yet non-trivial approach to analyze the field theory of collective degrees of freedom. For a detailed discussion of the higher order corrections in $1/N$, see [24].

3.3 Emergent Nematic Order in Iron Based Systems

This Section is based on the results of the following references [42, 65].

Emergent nematic order in iron based systems has been be explained by two different approaches – the orbital scenario [5, 74, 78, 82, 85, 86] and the spin-driven nematic scenario [42, 44, 45]. Both scenarios offer explanations for a variety of experimental observations, such as the softening of the elastic modulus, resistivity anisotropy, signatures in the Raman spectroscopy. For a detailed discussion see for example [44]. Here we follow the approach of a spin-driven nematic state.

3.3.1 Heuristic Picture of the Spin-Driven Nematic Scenario

Before going deeper into the technical details, we will first outline an intuitive physical picture of the spin-driven nematic scenario. The heuristic picture of spin-driven nematicity is simple and can be described using symmetry arguments, as shown in Fig. 3.1. The stripe spin-density wave (SDW) phase breaks the $O(3)$ spin-rotational symmetry. Because of its non-zero ordering vector, translational symmetry breaking also takes place, which leads to an increase in the size of the unit cell in the magnetically ordered phase. In the iron pnictides, however, the situation is more subtle. The SDW phase supports two different ground-states – one in which the spins are parallel along the y-axis, and anti-parallel along x (ordering vector $\mathbf{Q}_x = (\pi, 0)$), or vice versa (ordering vector $\mathbf{Q}_y = (\pi, 0)$). Therefore, in the SDW state both $O(3)$ and Z_2 symmetries are broken. By Z_2 we mean a discrete Ising-like symmetry, which corresponds to one of the two choices for the ordering wave vector $\mathbf{Q}_x$ or $\mathbf{Q}_y$. The Z_2 symmetry is expected to be less influenced by fluctuations than the $O(3)$ symmetry breaking, because it is a discrete symmetry and therefore not subject to Mermin-Wagner's theorem. Therefore, when one includes the fluctuations, the Z_2 symmetry

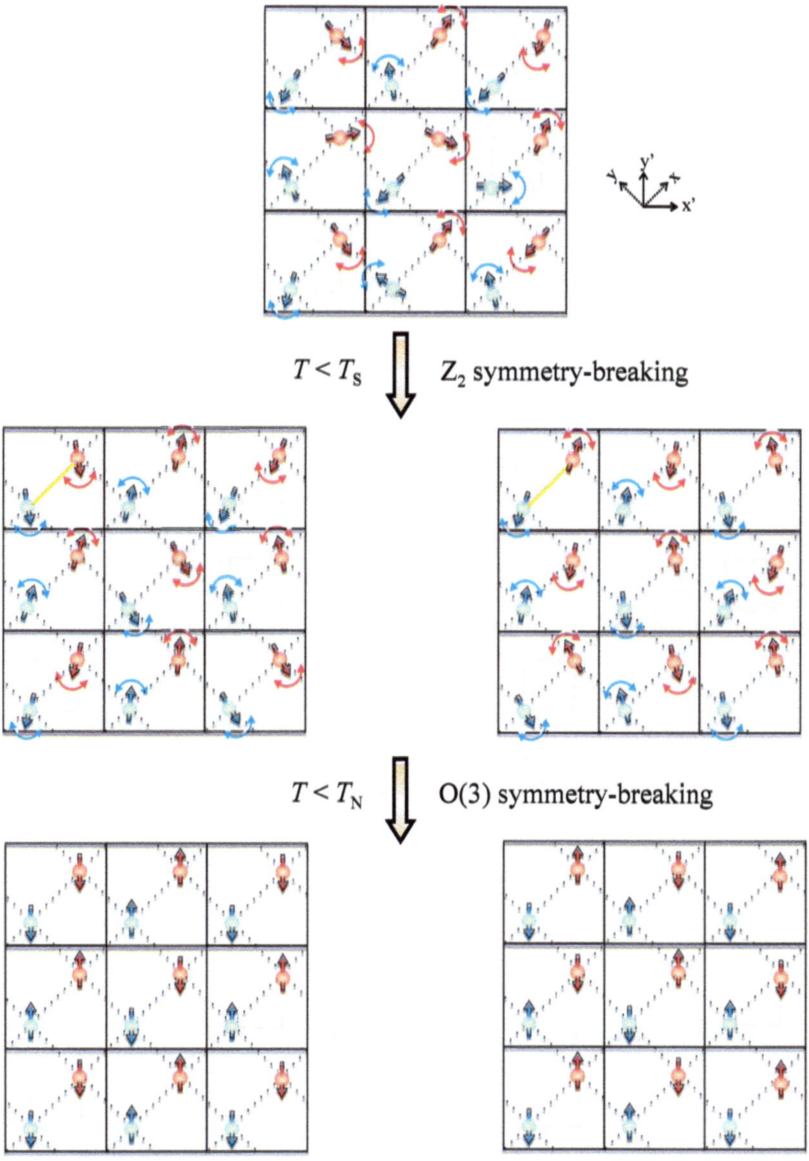

Fig. 3.1 Schematic representation of the nematic transition in real space. The stripe spin-density wave order parameter manifold is characterized by $O(3) \times Z_2$ symmetry. *Upper panel* in the tetragonal phase $T > T_s$ both the $O(3)$ and the Z_2 symmetry remain unbroken. *Middle panel* First, only the Z_2 symmetry is broken, while the $O(3)$ symmetry remains unbroken: the system is paramagnetic, since $\langle \mathbf{S}_i \rangle = 0$ (indicated by the double-arrow on top of the spins), but the spin correlations break the tetragonal symmetry, $\langle \mathbf{S}_i \cdot \mathbf{S}_{i+\mathbf{x}} \rangle = -\langle \mathbf{S}_i \cdot \mathbf{S}_{i+\mathbf{y}} \rangle$ (*yellow bonds*). This is the nematic phase with broken Z_2 (Ising) symmetry, but no magnetic order, which sets in at the temperature T_N. *Lower panel* In the second step, the $O(3)$ symmetry also gets broken and the system develops magnetic order, characterized by two degenerate ground states of magnetic stripes with parallel spins along the y axis (ordering vector $\mathbf{Q}_x = (\pi, 0)$) or along the x axis (ordering vector $\mathbf{Q}_y = (0, \pi)$)

can get broken before the $O(3)$ symmetry. In this way an intermediate phase, with the broken Z_2 symmetry but unbroken $O(3)$ symmetry can be stabilized. This is the Ising-nematic state in pnictides.

Note that the correlation functions between neighbouring spins $\langle \mathbf{S}_i \cdot \mathbf{S}_{i+\mathbf{x}} \rangle$ and $\langle \mathbf{S}_i \cdot \mathbf{S}_{i+\mathbf{y}} \rangle$ come with the opposite signs (see Fig. 3.1b). Therefore, in real space, the Z_2 symmetry-breaking corresponds to a broken tetragonal symmetry. Thus, although the translational and rotational symmetries are always broken in crystals, an additional lowering of the rotational symmetry occurs at the nematic transition. Finally, we stress that the driving force behind the nematic transition is electronic in nature [23], rather than this transition being an ordinary tetragonal-to-orthogonal structural transition.

In mean-field theory the Z_2 and the $O(3)$ symmetry breaking occurs simultaneously. However, either quantum or thermal fluctuations can split the transitions. In case of joint transitions, they are then both first order transitions [42]. The criterion for breaking the discrete Z_2 symmetry via a second order transition can be formulated in terms of a threshold value of the magnetic correlation length ξ. As one decreases the temperature and approaches the nematic transition, this leads to an increase of the magnetic correlation length. Before the magnetic correlation length starts diverging (which is the condition for a second order magnetic transition), the threshold value can be reached and spin-driven nematicity will set in. This naturally explains why the magnetic and structural phase boundaries are correlated. Let $\boldsymbol{\Delta}_{X/Y}$ denote the spin fluctuation field associated with the ordering along $\mathbf{Q}_{X/Y}$. We will further show that the nematic phase is characterized by an unequal strength of the magnetic fluctuations associated with the ordering wave vectors $\mathbf{Q}_X$ and $\mathbf{Q}_Y$: $\langle \boldsymbol{\Delta}_X^2 - \boldsymbol{\Delta}_Y^2 \rangle \neq 0$, but no long range magnetic order, $\langle \boldsymbol{\Delta}_{X,Y} \rangle = 0$. Therefore, it is the magnetic fluctuations that break the C_4 symmetry down to C_2.

Now, we will present a microscopic model of spin-driven nematicity, explicitly derive the condition for the onset of the nematic phase (by calculating the nematic susceptibility in the large N limit), and sketch the emerging phase diagram.

3.3.2 *Order-Parameter Theory of Stripe-density-wave Phase*

First, we will discuss the order parameter theory of striped magnetic order in iron-based superconductors, starting from a simplified model for the bandstructure, confirmed by experiments. We include the coupling to the lattice in all our calculations, as we will see later on how this can have crucial consequences on the nature of the nematic phase transition, and the signatures of nematicity in the Raman spectroscopy.

We will adopt a coordinate system of a single Fe unit cell, where the k_x and k_y axes are along the Fe-Fe bonds. In what follows we will present the mathematical model for the spin-driven nematic phase, summarizing the steps of [65]. We start from an itinerant model with the simplified band-structure presented in Fig. 3.2. We consider the bands near the $\Gamma-$ point and the $X-$ and $Y-$ points in the Brillouin zone. Our main results do not rely on the explicit knowledge of the detailed parametrization of

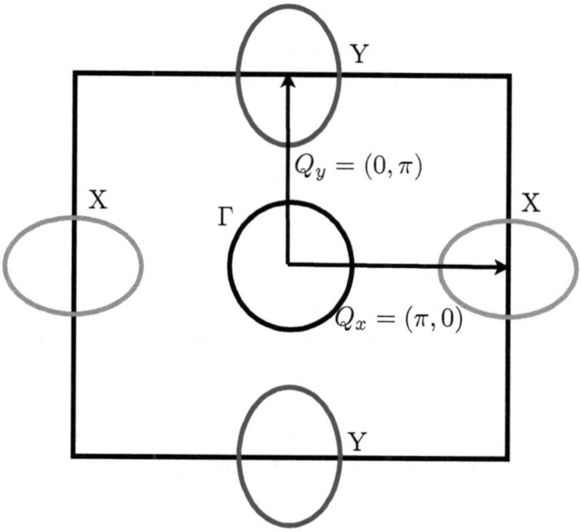

Fig. 3.2 Band structure: the central hole-like Γ band is circular, and the electron-like X (*red*) and Y (*blue*) bands, shifted by $\mathbf{Q}_X = (\pi, 0)$ and $\mathbf{Q}_Y = (0, \pi)$, respectively, are elliptical in our model (color figure online)

the band structure, except for the fact that the band-structure is not perfectly nested. The band dispersions are given by

$$\begin{aligned}
\epsilon_{\Gamma,\mathbf{k}} &= \epsilon_0 - \frac{k^2}{2m} - \mu, \\
\epsilon_{X,\mathbf{k}+\mathbf{Q_X}} &= -\epsilon_0 + \frac{k_x^2}{2m_x} + \frac{k_y^2}{2m_y} - \mu, \\
\epsilon_{Y,\mathbf{k}+\mathbf{Q_Y}} &= -\epsilon_0 + \frac{k_x^2}{2m_y} + \frac{k_y^2}{2m_x} - \mu,
\end{aligned} \tag{3.60}$$

where m_i are the band masses, ϵ_0 is the offset energy, and μ denotes the chemical potential. The corresponding Fermi surfaces are shown in Fig. 3.2.

We consider the Hamiltonian that contains interactions in the spin channel with momenta near $\mathbf{Q}_X$ and $\mathbf{Q}_Y$, as well as the elasto-nematic coupling:

$$\begin{aligned}
\mathcal{H} &= \mathcal{H}_0 + \mathcal{H}_{\text{int}} + \mathcal{H}_{\text{el-nem}}, \\
\mathcal{H}_0 &= \sum_{i,\mathbf{k}} \epsilon_{i,\mathbf{k}} c^\dagger_{i,\mathbf{k}\alpha} c_{i,\mathbf{k}\alpha}, \\
\mathcal{H}_{\text{int}} &= -\frac{1}{2} u_s \sum_{i,\mathbf{q}} \mathbf{s}_{i,\mathbf{q}} \cdot \mathbf{s}_{i,-\mathbf{q}}, \\
\mathcal{H}_{\text{el-nem}} &= \lambda_{\text{el}} \int \left(\mathbf{s}_X^2 - \mathbf{s}_Y^2\right) \left(\partial_x u_x - \partial_y u_y\right).
\end{aligned} \tag{3.61}$$

$c^{\dagger}_{i,\mathbf{k}\alpha}$ represents the operator that creates an electron with momentum $\mathbf{k}$, spin α and in the band i. In order to formulate a large-N theory, we consider $SU(\mathcal{N})$ spins, which leads to $N = \mathcal{N}^2 - 1$ component collective spin degrees of freedom. The spin operator is then given by

$$\mathbf{s}_{i,\mathbf{q}} = \sum_k c^{\dagger}_{\Gamma,\mathbf{k}+\mathbf{q}\alpha} \boldsymbol{\lambda}_{\alpha\beta} c_{i,\mathbf{k}\beta}. \tag{3.62}$$

Here $\boldsymbol{\lambda}_{\alpha\beta}$ is the $\mathcal{N}^2 - 1$ component vector of the generators of the $SU(\mathcal{N})$ algebra. For $\mathcal{N} = 2$ one has that $\boldsymbol{\lambda}_{\alpha\beta} = \frac{1}{2}\boldsymbol{\sigma}_{\alpha\beta}$ with vector of the Pauli matrixes $\boldsymbol{\sigma}$. u_s represents the coupling constant in the spin channel. This coupling constant is a combination of density-density and pair-hopping interactions between hole and electron pockets [25]. $\mathcal{H}_{\text{el-nem}}$ describes the nemato-elastic coupling Hamiltonian, since the nematic order parameter couples linearly to the orthorhombic distortion $\partial_x u_x - \partial_y u_y$, which plays an important role in Raman experiments, as has been shown in [53, 72]. λ_{el} is the nemato-elastic coupling constant and $\mathbf{u} = (u_x, u_y)$ the phonon displacement field. We define the spinor

$$\Psi^{\dagger}_k = \left(c^{\dagger}_{\Gamma,k,\alpha}\; c^{\dagger}_{X,k,\alpha}\; c^{\dagger}_{Y,k,\alpha} \right), \tag{3.63}$$

with α being the value of the $SU(\mathcal{N})$ spin index. The partition function then reads

$$\begin{aligned} Z &= \int d\Psi d\mathbf{u} e^{-S[\Psi,\mathbf{u}]}, \\ S[\Psi,\mathbf{u}] &= \int_0^{\beta} d\tau \int_x \left\{ \Psi^{\dagger}\partial_{\tau}\Psi + \mathcal{H}(\Psi^{\dagger},\Psi) \right\} + S_{ph}[\mathbf{u}], \\ S_{ph}[\mathbf{u}] &= \int \left[c_s^0 (\nabla\mathbf{u})^2 + (\partial_{\tau}\mathbf{u})^2 \right], \end{aligned} \tag{3.64}$$

where $\beta = T^{-1}$ is the inverse temperature, $S_{ph}[\mathbf{u}]$ the pure phononic action, and c_s^0 the bare value of the shear modulus.

Since, we are interested in analyzing the nematic phase which occurs as a consequence of spin fluctuations, we decouple the quartic interaction term (in fermionic fields) $\mathbf{s}.\mathbf{s}$ in $\mathcal{H}_{\text{int}}$ in (3.61) using a Hubbard–Stratonovich decoupling in the spin-channel. In this way, we eliminate the quartic interaction term between fermions in the partition function at the expense of having a functional integral over two additional bosonic fields $\boldsymbol{\Delta}_X$ and $\boldsymbol{\Delta}_Y$, with $\mathcal{N}^2 - 1$ components, similarly to how we introduced the bosonic field ϕ to perform the decoupling in (3.11). The bosonic fields couple linearly to the fermionic spin densities. The partition function is given by

$$Z = \int d\Delta_i d\Psi d\mathbf{u} e^{-S[\Psi,\Delta_i,\mathbf{u}]}, \tag{3.65}$$

with the following action:

$$S[\Psi, \Delta_i, \mathbf{u}] = -\int_k \Psi_k^\dagger \mathcal{G}_{\Delta,k}^{-1} \Psi_k + \frac{1}{2u_s}\int_x \left(\boldsymbol{\Delta}_X^2 + \boldsymbol{\Delta}_Y^2\right) + \frac{\lambda_{\rm el}}{u_s^2}\int \left(\boldsymbol{\Delta}_X^2 - \boldsymbol{\Delta}_Y^2\right)\left(\partial_x u_x - \partial_y u_y\right) + S_{ph}[\mathbf{u}]. \quad (3.66)$$

The matrix of the inverse Green's function $\mathcal{G}_k^{-1}$ reads:

$$\mathcal{G}_{\Delta,k}^{-1} = \mathcal{G}_{0,k}^{-1} - \mathcal{V}_\Delta, \quad (3.67)$$

with the bare Green function matrix:

$$\mathcal{G}_{0,k} = \begin{pmatrix} \hat{G}_{\Gamma,k} & 0 & 0 \\ 0 & \hat{G}_{X,k} & 0 \\ 0 & 0 & \hat{G}_{Y,k} \end{pmatrix}, \quad (3.68)$$

and the interacting term:

$$\mathcal{V}_\Delta = \begin{pmatrix} 0 & -\boldsymbol{\Delta}_X \cdot \boldsymbol{\lambda} & -\boldsymbol{\Delta}_Y \cdot \boldsymbol{\lambda} \\ -\boldsymbol{\Delta}_X \cdot \boldsymbol{\lambda} & 0 & 0 \\ -\boldsymbol{\Delta}_Y \cdot \boldsymbol{\lambda} & 0 & 0 \end{pmatrix}. \quad (3.69)$$

$\hat{G}_{i,k} = G_{i,k}\hat{1}$ with $G_{i,k}^{-1} = i\omega_n - \epsilon_{i,\mathbf{k}}$ and $\mathcal{N} \times \mathcal{N}$ unit matrix $\hat{1}$. We can invert the matrix equation (3.67) to obtain

$$\mathcal{G}_\Delta = \sum_{n=0}^{\infty} \left(\mathcal{G}_0 \mathcal{V}_\Delta\right)^n \mathcal{G}_0. \quad (3.70)$$

Next, we integrate out both the fermionic degrees of freedom and phonons from (3.65):

$$Z = \int d\Delta_i e^{-S_{\rm eff}[\boldsymbol{\Delta}_X, \boldsymbol{\Delta}_Y]}, \quad (3.71)$$

with action:

$$S_{\rm eff}[\boldsymbol{\Delta}_X, \boldsymbol{\Delta}_Y] = -\mathrm{Tr}\ln\left(1 - \mathcal{G}_0\mathcal{V}_\Delta\right) + \frac{1}{2u_s}\int_x \left(\boldsymbol{\Delta}_X^2 + \boldsymbol{\Delta}_Y^2\right) - \frac{1}{2}\int g_{\text{nem-el}}\left(\boldsymbol{\Delta}_X^2 - \boldsymbol{\Delta}_Y^2\right), \quad (3.72)$$

where $g_{\text{nem-el}}(q, \omega) = \frac{\gamma_{\rm el}^2 q^2}{c_s^0 q^2 - \omega^2}$ is frequency and momentum-dependent nematic coupling constant, and $\gamma_{\rm el}^2 = \frac{1}{2}\left(\frac{\lambda_{\rm el}}{u_s^2}\right)^2$. $\mathrm{Tr}(\cdots)$ refers to trace over momentum, frequency, spin, and band indices. We expand the action in powers of spin-fluctuation fields $\boldsymbol{\Delta}_{X,Y}$ to obtain:

$$S_{\text{eff}}\left[\boldsymbol{\Delta}_X, \boldsymbol{\Delta}_Y\right] = \frac{1}{2}\mathrm{Tr}\left(\mathscr{G}_{0,k}\mathscr{V}_{\Delta}\right)^2 + \frac{1}{4}\mathrm{Tr}\left(\mathscr{G}_{0,k}\mathscr{V}_{\Delta}\right)^4 + \frac{2}{u_{\mathrm{s}}}\int_x \left(\boldsymbol{\Delta}_X^2 + \boldsymbol{\Delta}_Y^2\right) \\ -\frac{1}{2}\int g_{\text{nem-el}}\left(\boldsymbol{\Delta}_X^2 - \boldsymbol{\Delta}_Y^2\right). \tag{3.73}$$

The traces over the $SU(\mathscr{N})$ indices in the above are rather tedious to evaluate for a general $\mathscr{N}$, and their evaluation will not be presented here. They were evaluated in the Appendix of [65], giving the following result for the effective action in the large $\mathscr{N}$ limit:

$$S_{\text{eff}}\left[\boldsymbol{\Delta}_X, \boldsymbol{\Delta}_Y\right] = \sum_i r_{0,i}\Delta_i^2 + \sum_{i,j} u_{ij}\Delta_i^2\Delta_j^2 - \frac{1}{2}\int g_{\text{nem-el}}\left(\boldsymbol{\Delta}_X^2 - \boldsymbol{\Delta}_Y^2\right) \tag{3.74}$$

with the coefficients:

$$r_{0,i} = \frac{1}{2u_{\mathrm{s}}} + \frac{1}{2}\int_k G_{\Gamma,k}G_{i,k}, \\ u_{ij} = \frac{1}{8\mathscr{N}}\int_k G_{\Gamma,k}^2 G_{i,k}G_{j,k}. \tag{3.75}$$

Due to the underlying C_4 symmetry which is present in the tetragonal phase the following identities hold

$$\int_k G_{\Gamma,k}G_{X,k} = \int_k G_{\Gamma,k}G_{Y,k}, \\ \int_k G_{\Gamma,k}^2 G_{X,k}^2 = \int_k G_{\Gamma,k}^2 G_{Y,k}^2. \tag{3.76}$$

The action can then be simplified to

$$S_{\text{eff}}[\boldsymbol{\Delta}_X, \boldsymbol{\Delta}_Y] = r_0(\boldsymbol{\Delta}_X^2 + \boldsymbol{\Delta}_Y^2) + \frac{u}{2}(\boldsymbol{\Delta}_X^2 + \boldsymbol{\Delta}_Y^2)^2 - \frac{g_r}{2}(\boldsymbol{\Delta}_X^2 - \boldsymbol{\Delta}_Y^2)^2, \tag{3.77}$$

with coefficients

$$r_0 = \frac{1}{2u_s} + \frac{1}{2}\int G_{X,k}G_{\Gamma,k}, \\ u = \frac{1}{16\mathscr{N}}\int_k G_{\Gamma,k}^2(G_{X,k} + G_{Y,k})^2, \\ g_r(q,\omega) = g + \gamma_{\text{el}}^2\frac{q^2}{c_{\mathrm{s}}^0 q^2 - \omega^2}, \\ g = -\frac{1}{16\mathscr{N}}\int_k G_{\Gamma,k}^2(G_{X,k} - G_{Y,k})^2. \tag{3.78}$$

r_0 tunes the distance from the magnetic transition and the interaction constants u and g were calculated as a function of temperature and ellipticity in [42]. It was found that the coupling g vanishes for circular electron pockets, but is positive for a non-zero ellipticity.

3.3.3 *Ising-Nematic Order*

3.3.3.1 Effective Action in Terms of the Nematic Degrees of Freedom

Next, we decouple the quartic terms in the action (3.77) by introducing two Hubbard–Stratonovich scalar fields: $\varphi \to \boldsymbol{\Delta}_X^2 - \boldsymbol{\Delta}_Y^2$ and $i\lambda \to \boldsymbol{\Delta}_X^2 + \boldsymbol{\Delta}_Y^2$. After the decoupling, the effective action reads

$$S_{\text{eff}} = \int_q \chi_q^{-1} \left(\boldsymbol{\Delta}_X^2 + \boldsymbol{\Delta}_Y^2\right) + \int \left(\frac{\varphi^2}{2g_r} + \frac{\lambda^2}{2u}\right) + \int_x i\lambda \left(\boldsymbol{\Delta}_X^2 + \boldsymbol{\Delta}_Y^2\right) + \int_x (\varphi + h_{\text{n}}) \left(\boldsymbol{\Delta}_X^2 - \boldsymbol{\Delta}_Y^2\right), \tag{3.79}$$

where we have added a field h_{n} conjugate to the nematic order parameter $\Delta_X^2 - \Delta_Y^2$. This term will be used to calculate the susceptibility of the nematic order parameter. A finite value of φ results in non-zero value of $\frac{\varphi}{g_r} = \langle \boldsymbol{\Delta}_X^2 - \boldsymbol{\Delta}_Y^2 \rangle \neq 0$ and the system develops a nematic phase. The field λ always has a non-zero large-N mean field value and characterizes the strength of magnetic fluctuations. Next we want to investigate the possibility of the split magnetic and nematic transition, such that there is no magnetic ordering right below the structural transition temperature in the nematic phase, i.e. $\langle \boldsymbol{\Delta}_{X,Y} \rangle = 0$. Therefore, we are allowed to integrate over the $\mathcal{N}^2 - 1$ component fields $\boldsymbol{\Delta}_{X,Y}$:

$$S_{\text{eff}}[\lambda, \varphi] = N \int_q \left\{\frac{\varphi^2}{2g_r} + \frac{\lambda^2}{2u}\right\} + \frac{N}{2} \int_q \left\{\log\left[\left(\chi_q^{-1} + i\lambda\right)^2 - (\varphi + h_{\text{n}})^2\right]\right\}. \tag{3.80}$$

3.3.3.2 Large-N Equation of State

The effective action (3.80) comes with an overall pre-factor of N.

In the limit $N \gg 1$, we analyze the action via the saddle-point method, similar to the spirit of Sect. 3.2.2. In [42] a comparison of large-N expansion with the renormalization group method was performed and very similar results for the phase diagram were obtained at the upper critical dimension, which justifies the use of the large N expansion in this case. The extremum of the action is given by $\partial S_{\text{eff}} [\varphi, \lambda] / \partial \varphi = \partial S_{\text{eff}} [\varphi, \lambda] / \partial \lambda = 0$, and we obtain the equations of state for φ

and λ:

$$\frac{i\lambda}{u} = \int_q \frac{\chi_q^{-1} + i\lambda}{\left(\chi_q^{-1} + i\lambda\right)^2 - (\varphi + h_{\rm n})^2},$$
$$\frac{\varphi}{g_r} = \int_q \frac{\varphi + h_{\rm n}}{\left(\chi_q^{-1} + i\lambda\right)^2 - (\varphi + h_{\rm n})^2}. \qquad (3.81)$$

3.3.3.3 Nematic Susceptibility in the Large-N Expansion

We differentiate the second equation in (3.81) with respect to the conjugate field, in order to find the static nematic susceptibility

$$\chi_{\rm nem} = \left.\frac{\partial\varphi}{\partial h_{\rm n}}\right|_{h_{\rm n}=0} = \frac{g_{\rm stat}\int_k \chi_k^2}{1 - g_{\rm stat}\int_k \chi_k^2}, \qquad (3.82)$$

where we have shifted $\chi_k^{-1} \to \chi_k^{-1} + i\lambda$, which corresponds to the re-normalisation of the magnetic transition temperature due to fluctuations, and where

$$g_{\rm stat} = g + \frac{\gamma_{\rm el}^2}{c_{\rm s}^0} \qquad (3.83)$$

is the static limit ($\omega = 0$) of $g_r(q, \omega)$, given by (3.78).

3.3.4 *Sketch of the Phase Diagram*

Now that we derived the expression for the nematic susceptibility, (3.82), we are in the position to analyze the onset of the nematic phase. The condition for a second order nematic transition to occur is that $\chi_{\rm nem} \to \infty$. This happens at the finite value of the magnetic correlation length ξ, which is given by $r = r_0 + i\lambda = \xi^{-2}$, when the condition $g_{\rm stat}\int_k \chi_k^2 = 1$ is satisfied. As a result, the nematic phase pre-empts the magnetic phase, and occurs as a thin sliver in the phase diagram, just above the magnetic transition temperature, for details see Fig. 3.3. The splitting between the nematic and magnetic transition is controlled by the nematic coupling constant $g_{\rm stat}$ in our model, which in turns depends on various parameters like temperature, detailed band-structure, and magnitude of the nemato-elastic coupling. Changing the doping or pressure (as showed in the phase diagram of Fig. 3.3) will effectively change $g_{\rm stat}$ in our model and control the size of the splitting between the magnetic and nematic transitions. It has been shown that a first order nematic transition can occur as well in systems with moderate out-of-plane spin anisotropy [42].

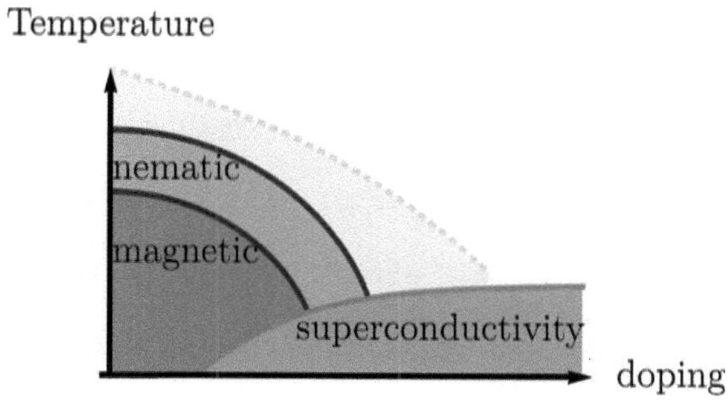

Fig. 3.3 Phase diagram of the iron pnictides. The nematic phase is colored *blue* and occurs as a region in the phase diagram above the magnetic transition temperature, closely tracking the magnetic phase-boundary. The area to which nematic fluctuations extend is *shaded light blue*. The dashed line is not a phase boundary, but a crossover temperature where nematic fluctuations become sizable

Indeed, there is a vast number of experimental examples where it has been shown that the nematic transition at T_s (the onset of nematic order coincides with the structural transition temperature T_s) closely tracks the magnetic transition at T_N [45, 66, 69, 98], i.e.: $T_s \geq T_N$. The 1111 materials support split structural and magnetic phase transitions which are of second order. On the other hand in the 122 family, the magnetic and nematic transition are either joint and of first order or split. In $Ba(Fe_{1-x}Co_x)_2As_2$, where the transitions are split, the magnetic transition is weakly first order for $x < 0.02$ [69, 98].

In what follows, we will concentrate on the case of a second order nematic transition, and a second order magnetic transition.

3.4 Elastic Coupling and Spin-Driven Nematicity in Iron-Based Superconductors

This Section is based upon the results of [64].

Next, we would like to develop the φ^4-theory for the nematic degrees of freedom, in order to analyze the nematic fluctuations and the nematic transition, using renormalization group (RG) analysis. We are motivated by the experiments [10, 11, 113] where the Curie-Weiss behaviour of the nematic susceptibility ($\chi_{\text{nem}} \sim \frac{1}{T-T_s}$) was observed in a rather large temperature window, suggesting the mean-field behavior of nematic degrees of freedom.

Before we begin to analyze the effects of the elastic coupling on the nematic degrees of freedom in detail, we will present a short summary of the elastic theory, and derive the dispersion of the acoustic mode in a system with tetragonal symmetry, relevant for iron-based superconductors.

3.4.1 Elastic Theory of a Tetragonal System

The elastic part of the free energy of a tetragonal system is given by

$$\begin{aligned}F_{\text{el}} = & \tfrac{c_{11}}{2}\left(\epsilon_{xx}^2+\epsilon_{yy}^2\right)+\tfrac{c_{33}}{2}\epsilon_{zz}^2+\tfrac{c_{44}}{2}\left(\epsilon_{xz}^2+\epsilon_{yz}^2\right)\\ & +\tfrac{c_{66}}{2}\epsilon_{xy}^2+c_{12}\epsilon_{xx}\epsilon_{yy}+c_{13}\left(\epsilon_{xx}+\epsilon_{yy}\right)\epsilon_{zz},\end{aligned} \tag{3.84}$$

where $\epsilon_{ij}=\frac{\partial_i u_j+\partial_j u_i}{2}$ is the strain tensor, and u_i is the ith component of the phonon displacement field.

The dynamic matrix M can be deduced from $F_{\text{el}}=\frac{1}{2}u_i(q)M_{ij}(q)u_j(q)$. It then follows that

$$M_{ij}=\sum_{m,l}q_m q_l\frac{\partial^2 F_{\text{el}}}{\partial\epsilon_{im}\partial\epsilon_{lj}}. \tag{3.85}$$

From (3.84) and (3.85) it then follows that for a tetragonal system the dynamic matrix reads

$$M(\mathbf{q})=\begin{pmatrix} c_{11}q_x^2+c_{66}q_y^2+c_{44}q_z^2 & (c_{12}+c_{66})\,q_xq_y & (c_{13}+c_{44})\,q_xq_z\\ (c_{12}+c_{66})\,q_xq_y & c_{66}q_x^2+c_{11}q_y^2+c_{44}q_z^2 & (c_{13}+c_{44})\,q_yq_z\\ (c_{13}+c_{44})\,q_xq_z & (c_{13}+c_{44})\,q_yq_z & c_{44}\left(q_x^2+q_y^2\right)+c_{33}q_z^2\end{pmatrix}. \tag{3.86}$$

The dynamic matrix M determines the phonon dispersion ω via $\det\left(\omega^2\rho-M(\mathbf{q})\right)=0$, where ρ denotes the density. A vanishing elastic constant results in a vanishing sound velocity. The transition from a tetragonal state into an orthorhombic state is characterized by a vanishing elastic constant $c_{11}-c_{12}\to 0$ in the one Fe unit cell system (It is a vanishing c_{66} in the two Fe unit cell).

The directions along which the sound velocity vanishes are given by $q_z=0$ and $q_x=\pm q_y$. Along these directions it holds that $\omega^2\rho=(c_{11}-c_{12})q_x^2\to 0$. The dispersion in the vicinity of such line $q_x=q_y$ is given by

$$\omega^2\rho\approx(c_{11}-c_{12})\frac{\left(q_x+q_y\right)^2}{2}+\frac{\left(q_x-q_y\right)^2}{2}\left[c_{11}+c_{66}-\frac{\left(c_{11}-c_{66}\right)^2}{c_{12}+c_{66}}\right]+c_{44}q_z^2, \tag{3.87}$$

where $q_z\ll q_x$ and $q_x-q_y\ll q_x$. We introduce the angles ϕ and θ through $q_x=q\sin\theta\cos\left(\phi+\frac{\pi}{4}\right)$, and $q_y=q\sin\theta\sin\left(\phi+\frac{\pi}{4}\right)$, and get that $\omega^2\rho\approx c_s(\mathbf{q})q_+^2$, with $q_+=(q_x+q_y)$ being the soft momentum and

$$c_s(\mathbf{q})=\frac{(c_{11}-c_{12})}{2}+\mu_1\sin^4\theta\sin^2 2\phi+\mu_2\cos^2\theta, \tag{3.88}$$

with

$$\mu_1 = \frac{1}{8}\left[c_{11} + c_{66} - \frac{(c_{11} - c_{66})^2}{c_{12} + c_{66}}\right],$$
$$\mu_2 = c_{44}. \tag{3.89}$$

3.4.2 Model

We start from the action derived in Sect. 3.3.2, but this time we explicitly include the momentum dependence of the elastic constant $c_s(\mathbf{q})$, given in (3.88) in the phononic action. As we will show, this will have crucial consequences for the nature of the nematic transition.

The action is given by

$$S = S_\Delta + S_{\text{ph}} + S_c, \tag{3.90}$$

with

$$Z = \int \mathcal{D}\mathbf{u}\mathcal{D}\left(\boldsymbol{\Delta}_x\boldsymbol{\Delta}_y\right) e^{-S}. \tag{3.91}$$

Here, S_Δ represents the magnetic part of the action, S_{ph} the phononic action, and S_c the coupling between lattice and magnetic degrees of freedom. The magnetic action, derived in Sect. 3.3.2, see (3.77) is given by

$$S_\Delta = \int_x \left[r_0\left(\boldsymbol{\Delta}_x^2 + \boldsymbol{\Delta}_y^2\right) + (\nabla\boldsymbol{\Delta}_x)^2 + \left(\nabla\boldsymbol{\Delta}_y\right)^2\right]$$
$$+ \int_x \left[\frac{u}{2N}\left(\boldsymbol{\Delta}_x^2 + \boldsymbol{\Delta}_y^2\right)^2 - \frac{g}{2N}\left(\boldsymbol{\Delta}_x^2 - \boldsymbol{\Delta}_y^2\right)^2\right]. \tag{3.92}$$

In what follows we will pursue a continuum's model with variable dimensionality $2 < d < 3$. Such a model behaves in a very similar way to an anisotropic system of moderately or weakly coupled layers [42], which is a more-realistic description of the iron-based superconductors.

For our purpose it is enough to consider only the static ($\omega = 0$) part of the phonon action (3.64) in the B_{1g}-channel:

$$S_{\text{ph}} = N\int_q c_s(\mathbf{q})\left(q_x u_x - q_y u_y\right)^2, \tag{3.93}$$

where the momentum dependence of $c_s(\mathbf{q})$, given by (3.88) has been included explicitly. The coefficient N in S_{ph} was introduced to generate a consistent large N expansion.

As before, see (3.66), the magneto elastic coupling is

$$S_c = \lambda_{\rm el} \int_x \left(\boldsymbol{\Delta}_x^2 - \boldsymbol{\Delta}_y^2\right)\left(\partial_x u_x - \partial_y u_y\right). \tag{3.94}$$

3.4.3 φ^4 Theory of Nematic Degrees of Freedom

We start from the (3.79), conveniently written out in the matrix form

$$S_\Delta[\varphi,\lambda] = \frac{1}{2}\int_x \left(\frac{N}{u}\lambda^2 + \frac{N}{g}\varphi^2\right) + \int_x \begin{pmatrix}\boldsymbol{\Delta}_x \\ \boldsymbol{\Delta}_y\end{pmatrix}^T \mathscr{G}^{-1}[\lambda,\varphi]\begin{pmatrix}\boldsymbol{\Delta}_x \\ \boldsymbol{\Delta}_y\end{pmatrix}, \tag{3.95}$$

where

$$\mathscr{G}^{-1}[\lambda,\varphi] = \begin{pmatrix} r_0 + i\lambda + \varphi - \nabla^2 & 0 \\ 0 & r_0 + i\lambda - \varphi - \nabla^2 \end{pmatrix}. \tag{3.96}$$

We shift $\varphi \to \varphi + \lambda_{\rm el}\left(\partial_x u_x - \partial_y u_y\right)$ and integrate out the magnetic degrees of freedom. The action is then given by

$$\begin{aligned} S = &\frac{N}{2}\int_x \left(\frac{1}{u}\lambda^2 + \frac{1}{g}\varphi^2\right) + N\int_q c_s(\mathbf{q})\left(q_x u_x - q_y u_y\right)^2 \\ &- \frac{\lambda_{\rm el} N}{g}\int_x \varphi\left(\partial_x u_x - \partial_y u_y\right) + \frac{N}{2}\mathrm{tr}\log \mathscr{G}^{-1}[\lambda,\varphi], \end{aligned} \tag{3.97}$$

with $c_s(\mathbf{q})$ given by (3.88) and we have shifted $c_s \to c_s + \lambda_{\rm el}^2/g$. Integrating out the phonon degrees of freedom gives

$$S = \frac{N}{2}\int_q \left(\frac{1}{u}\lambda_q\lambda_{-q} + \left(\frac{1}{g} + \frac{\lambda_{\rm el}^2}{c_s(\mathbf{q})}\right)\varphi_q\varphi_{-q}\right) + \frac{N}{2}\mathrm{tr}\log\mathscr{G}^{-1}[\lambda,\varphi]. \tag{3.98}$$

In what follows, we will focus on the tetragonal phase. We write

$$\lambda(x) = -i\psi_0 + \eta(x), \tag{3.99}$$

where ψ_0 is the saddle point solution of the large N expansion and amounts to a fluctuation renormalization of the magnetic transition temperature. $\eta(x)$ denotes the fluctuating part of $\lambda(x)$. The Green's function matrix is

$$\mathscr{G}^{-1}[\lambda,\varphi] = \mathscr{G}_0^{-1} - \mathscr{V}[\eta,\varphi], \tag{3.100}$$

where

$$\mathscr{G}_0^{-1} = \left(r - \nabla^2\right) I, \tag{3.101}$$

and $r = r_0 + \psi_0$, with

$$\mathscr{V}[\eta,\varphi] = -\begin{pmatrix} i\eta+\varphi & 0 \\ 0 & i\eta-\varphi \end{pmatrix}. \tag{3.102}$$

We further expand:

$$\begin{aligned} \mathrm{tr}\log\mathscr{G}^{-1}[\lambda,\varphi] &= \mathrm{tr}\log\mathscr{G}_0^{-1} + \mathrm{tr}\log(1-\mathscr{G}_0\mathscr{V}) \\ &\approx \mathrm{tr}\log\mathscr{G}_0^{-1} - \frac{1}{2}\mathrm{tr}(\mathscr{G}_0\mathscr{V})^2 - \frac{1}{3}\mathrm{tr}(\mathscr{G}_0\mathscr{V})^3 - \frac{1}{4}\mathrm{tr}(\mathscr{G}_0\mathscr{V})^4. \end{aligned} \tag{3.103}$$

The total action is given as a sum $S = S_0 + S_2 + S_3 + S_4$, where

$$S_0 = -\frac{N}{2}\int_x \frac{\psi_0^2}{u} + \frac{N}{2}\mathrm{tr}\log\mathscr{G}_0^{-1}, \tag{3.104}$$

and the quadratic part of the action reads

$$S_2 = \frac{N}{2}\int_q \begin{pmatrix} \eta(q) \\ \varphi(q) \end{pmatrix}^T D^{-1}(q) \begin{pmatrix} \eta(-q) \\ \varphi(-q) \end{pmatrix}, \tag{3.105}$$

with

$$D^{-1}(q) = \begin{pmatrix} D_\eta^{-1}(q) & 0 \\ 0 & D_\varphi^{-1}(q) \end{pmatrix}, \tag{3.106}$$

where $D_\eta^{-1}(q) = \frac{1}{u} + \Pi(q,r)$ and $D_\varphi^{-1}(q) = \frac{1}{g} + \frac{\lambda^2}{c_s(q)} - \Pi(q,r)$. The self-energy part is

$$\begin{aligned} \Pi(q,r) &= \int_k \frac{1}{r+k^2}\frac{1}{r+(\mathbf{k}+\mathbf{q})^2} \\ &\approx L_2 r^{\frac{d}{2}-2} - b_\varphi q^2, \end{aligned} \tag{3.107}$$

with

$$b_\varphi = -r^{\frac{d}{2}-3}\left(\frac{4}{d}L_4 + \frac{d-4}{d}L_3\right), \tag{3.108}$$

and we have defined

$$L_n = \int_p \frac{1}{\left(1+p^2\right)^n} = \frac{\Gamma\left(n-\frac{d}{2}\right)}{\left(2\sqrt{\pi}\right)^d\,\Gamma(n)}. \tag{3.109}$$

The cubic term in the action reads

$$S_3 = -N\int_{p,q} T(p,q)\varphi(-p)\varphi(p-q)\eta(q), \tag{3.110}$$

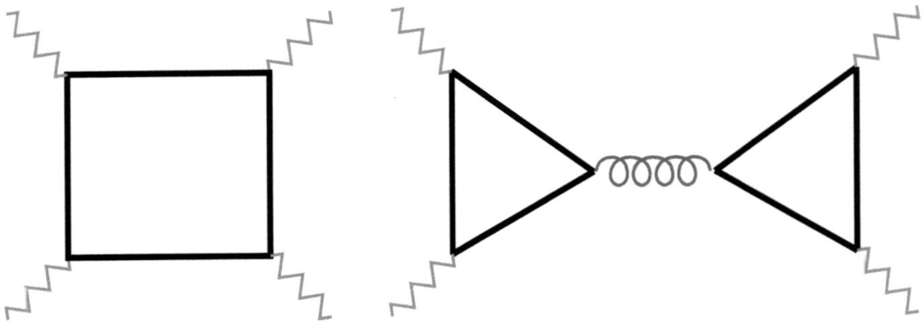

Fig. 3.4 Diagrammatic representation of the quartic coefficient u_φ. Spin-fluctuation propagators G_0 are denoted by full lines, while sharp wavy lines (*green*) denote the propagators D_φ. Spiral lines (*red*) represent the propagator D_η (see the main text). The left diagram is proportional to L_4 and gives a negative contribution to u_φ. The right diagram contains two triangular loops T given by (3.111), connected by a propagator D_η. It becomes larger for low dimensionality d and it provides a positive contribution to u_φ Figure adapted from [64]

where

$$T(p,q) = i\int_k G_0(k)G_0(k-q)G_0(p+k-q) \tag{3.111}$$

is the triangle depicted in Fig. 3.4.

The quartic terms in the action are given by

$$S_4 = -\frac{N}{4}L_4 r^{d/2-4}\int_{q_1,q_2,q_3}\varphi_{q_1}\varphi_{q_2}\varphi_{q_3}\varphi_{-q_1-q_2-q_3}. \tag{3.112}$$

Because the total action S is quadratic in η, we integrate out the η-fields from the action, using the following identity

$$\int \mathcal{D}\eta \exp\left\{\int_q\left[-\frac{N}{2}\eta(q)D_\eta^{-1}(q)\eta(-q) + Nj(q)\eta(q)\right]\right\}$$
$$= \exp\left\{-\frac{N}{2}j(q)D_\eta(q)j(-q)\right\}, \tag{3.113}$$

with $j(q) = \int_p T(p,q)\varphi(-p)\varphi(p-q)$.

The resulting action for the nematic fluctuations is therefore

$$S_\varphi/N = \frac{1}{2}\int_q\left(r_\varphi + \frac{\lambda_{\text{el}}^2}{c_s(\mathbf{q})} + b_\varphi q^2\right)\varphi_q\varphi_{-q} + \frac{u_\varphi}{4}\int_x \varphi^4, \tag{3.114}$$

where

$$\begin{aligned}
u_\varphi &= -L_4 r^{\frac{d}{2}-4} + \frac{2uL_3^2 r^{d-6}}{1+uL_2 r^{\frac{d}{2}-2}},\\
r_\varphi &= \frac{1}{g} - L_2 r^{\frac{d}{2}-2},\\
b_\varphi &= r^{\frac{d}{2}-3}\left(\frac{4}{d}L_4 + \frac{d-4}{d}L_3\right).
\end{aligned} \tag{3.115}$$

There are two competing contributions to the quartic term u_φ. The first term, proportional to L_4 is mean-field like and gives a negative contribution. The second term, proportional to L_3^2 contains two triangular loops T connected by the η propagator, see Fig. 3.4. It yields a positive contribution to u_φ, and is larger for a lower dimensionality d. This term is responsible for the change in sign of the quartic coupling constant and for the possibility of second order split magnetic and nematic transitions. For $d < 4$, $r^{\frac{d}{2}-2} \gg 1$ as $r \to 0$, and we get that for a sufficiently large magnetic correlation length ξ

$$u_\varphi = \left(\frac{2L_3^3}{L_2} - L_4\right) r^{\frac{d}{2}-4}. \tag{3.116}$$

There are two different correlation lengths in our problem – (i) the magnetic correlation length ξ given by $r \propto \xi^{-2}$, which diverges at T_N, and (ii) the nematic correlation length ξ_φ given by $r_\varphi \propto \xi_\varphi^{-2}$, which diverges at the structural transition temperature T_s. The quartic coupling constant u_φ, given by (3.116), can only be positive for $d < 3$. Therefore, only for very large correlation length and $d < 3$, it is possible to have a second order transition.

The second order nematic transition occurs when $\tilde{r}_\varphi = r_\varphi + \frac{\lambda_{\text{el}}^2}{c_s^0} = 0$, with r_φ given by (3.115). This gives

$$\frac{1}{g} - L_2 r^{\frac{d}{2}-2} + \frac{\lambda_{\text{el}}^2}{c_s^0} = 0, \tag{3.117}$$

which occurs at the large (but finite) value of the magnetic correlation length $\xi \propto r^{-\frac{1}{2}}$. Therefore, the magnetic and nematic transitions are split and the nematic transition pre-empts the magnetic transition, occurring at a slightly higher temperature $T_s > T_N$, as previously found in Sect. 3.3.4. The temperature difference is dictated by the bare value of the nematic coupling constant g and the size of the nemato-elastic coupling constant.

3.4.3.1 The Case with No Coupling to the Strain

We first analyze the case without coupling to the strain, i.e. when $\lambda_{\text{el}} = 0$. We will show that in this limit, nematic fluctuations are expected to be strong. We first determine the location of the tri-critical point in the phase diagram. This is the point where $u_\varphi = 0$, at which both the line of second order phase transitions and first order phase transitions would end. For a second order nematic phase transition to occur, the condition $u_\varphi > 0$ must be fulfilled. At the nematic transition and for $\lambda_{el} = 0$ holds that $r_\varphi = 0$, i.e. $L_2 r^{\frac{d}{2}-2} = \frac{1}{g}$, which we solve for $r(g)$:

$$r(g) = \left(\frac{2^d \pi^{d/2}}{\Gamma\left(2 - \frac{d}{2}\right) g} \right)^{\frac{2}{d-4}}. \tag{3.118}$$

We use this to express r in (3.115) in terms of g and get that, with $\alpha = u/g$:

$$u_\varphi = \frac{4-d}{24 g^{\frac{8-d}{4-d}}} \frac{2\alpha(3-d) - (6-d)}{1+\alpha} \left(\frac{2^d \pi^{d/2}}{\Gamma\left(2 - \frac{d}{2}\right)} \right)^{\frac{4}{4-d}}. \tag{3.119}$$

From the above we see that in $d < 3$ one obtains a second order transition for

$$\alpha > \alpha_c = \frac{6-d}{2(3-d)}. \tag{3.120}$$

This agrees with the result of [42].

We now want to estimate the strength of nematic fluctuations. In particular, we will estimate the size of the Ginzburg regime – the regime of strong critical fluctuations. One would expect such critical nematic fluctuations for $d \leq 4$, and if $u_\varphi > 0$. In order to do so, we rescale the action in terms of natural dimensionless coupling constant $\hat{u}_\varphi$. We substitute $\varphi = \mu\phi$ and $q = \gamma k$, where $\gamma^2 = r_\varphi / b_\varphi$ and $\mu = b_\varphi^{\frac{d}{4}} r_\varphi^{-\frac{d+2}{4}}$ such that

$$S = \frac{1}{2} \int_k \left(1 + k^2\right) \phi_k \phi_{-k} + \frac{\hat{u}}{4} \int_{k_{1,2,3}} \phi_{k_1} \phi_{k_2} \phi_{k_3} \phi_{-k_1-k_2-k_3}, \tag{3.121}$$

with

$$\begin{aligned} \hat{u} &= u_\varphi \gamma^{3d} \mu^4 \\ &= \frac{u_\varphi r_\varphi^{\frac{d-4}{2}}}{b_\varphi^{d/2}}. \end{aligned} \tag{3.122}$$

We obtain:

$$\hat{u}_\varphi = \Upsilon_d \left(g r_\varphi\right)^{\frac{d-4}{2}}, \tag{3.123}$$

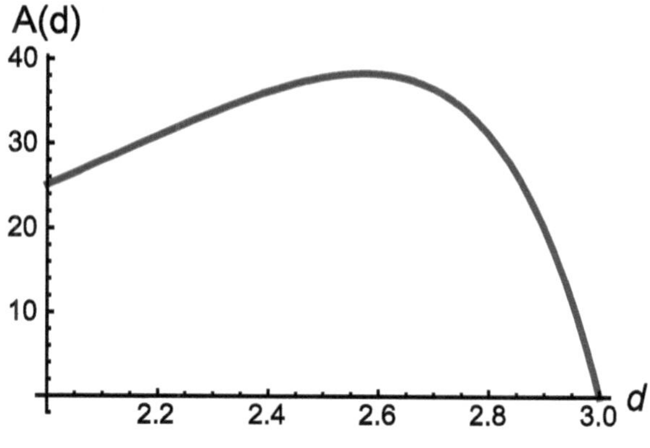

Fig. 3.5 Slope $A(d)$ of the coefficient Υ_d, near α_c, for $2 < d < 3$. The dimensionality $2 < d < 3$ effectively describes 3-dimensional, but anisotropic system, see [42]. The slope is always bigger than 1, except in the region around $d = 3$. This signifies strong fluctuation regime (except near $d = 3$) Figure adapted from [64]

where

$$\Upsilon_d = \frac{(48\pi)^{d/2}(4-d)^{1-\frac{d}{2}}}{24\Gamma\left(2-\frac{d}{2}\right)} \frac{2\alpha(3-d)-(6-d)}{(1+\alpha)}. \tag{3.124}$$

We introduce $f = gr_\varphi$. If $f \to 0$ the coupling $\hat{u}_\varphi$ will diverge. The Ginzburg regime f_{Ginz} is dictated by the condition $\hat{u}_\varphi(f) \approx 1$. We examine the coefficient $\Upsilon_d(\alpha)$. Close to the tricritical point ($\alpha \approx \alpha_c$), $\Upsilon_d(\alpha)$ we have that $\Upsilon_d(\alpha) = A(d)\frac{\alpha-\alpha_c}{\alpha_c}$, where $A(d)$ is the slope.

The slope $A(d)$ is plotted in Fig. 3.5. It is alway bigger than 1, except in the vicinity of $d = 3$. This means that in the case without coupling to the lattice the nematic fluctuations are rather strong, except in the immediate vicinity of the tricritical point.

3.4.3.2 The Case with Coupling to the Strain

We now include the coupling to the strain, to see if this will alter the conclusions of Sect. 3.4.3.1. We analyze the action S_φ of (3.114) with finite λ_{el}.

The action of a magnetic system in the presence of dipolar interactions closely resembles the action of (3.114) – i.e. both have the same structural form. This action was already analyzed in the 70 s using the renormalization group (RG) approach in [2, 3], in the context of uniaxial magnets with dipolar interactions in d dimensions. There, it was noted that the dipolar interactions generate non-analytic directional-dependent terms in the propagator for spin-fluctuations. Due to the existence of these non-analytic terms, the upper critical dimension is lowered from $d = 4$ (for the model without any dipolar interactions, with only exchange interaction) to $d = 3$ (for the model with the dipolar interaction).

The action for the nematic degrees of freedom (3.114), with the elastic modulus given in (3.88) contains similar non-analytic terms in the propagator for nematic fluctuations. In particular, we note that in the presence of the coupling to the lattice, only the lines $q_x = \pm q_y$, $q_z = 0$ remain soft, see Sect. 3.4.1 for more details.

We note that the nematic order parameter couples linearly to orthorhombic distortion, see (3.94). It is a well-known fact that the fluctuations of such an order parameter become suppressed [29, 46, 47, 80] and that this can lead to mean-field behaviour [29, 46, 80]. For iron based systems, this effect was already pointed out by Cano et al. [16], and in [114].

We now explicitly show, using RG analysis, that the upper critical dimension gets reduced in the presence of the coupling to the lattice. We start from the action of (3.114) and rescale the field to eliminate the coefficient b_φ:

$$S_\varphi = \frac{1}{2}\int_q \left(r_\varphi + \frac{\lambda_{\text{el}}^2}{c_s(q)} + q^2\right)\varphi_q\varphi_{-q} + \frac{u_\varphi}{4}\int_{q_1,q_2,q_3}\varphi_{q_1}\varphi_{q_2}\varphi_{q_3}\varphi_{-q_1-q_2-q_3}. \quad (3.125)$$

We define $\tilde{r}_\varphi = r_\varphi + \frac{\lambda_{\text{el}}^2}{c_s^0}$. The RG flow equations for this action are straightforward to derive, and we obtain (see for example [20] for the RG derivation for a φ^4 theory):

$$\begin{aligned}\frac{d\tilde{r}_\varphi}{dl} &= 2\tilde{r}_\varphi + 3u_\varphi\frac{d}{dl}\int_q^> D_\varphi(q),\\ \frac{du_\varphi}{dl} &= (4-d)\,u_\varphi - 9u_\varphi^2\frac{d}{dl}\int_q^> D_\varphi^2(q).\end{aligned} \quad (3.126)$$

The momentum integral runs over momenta between $\Lambda/b < q < \Lambda$, where Λ is the cut-off and $b = e^{-l}$. The nematic propagator is given by

$$D_\varphi^{-1}(q) \approx \tilde{r}_\varphi + q^2 + h_2\Lambda^2\sin^4\theta\sin^2 2\phi + h_1\Lambda^2\cos^2\theta, \quad (3.127)$$

where we defined $h_i\Lambda^2 = -\frac{\lambda^2}{(c_s^0)^2}\mu_i$. Power counting arguments show that the coupling constants h_i are relevant and they grow like

$$\frac{dh_i}{dl} = 2h_i. \quad (3.128)$$

Interaction corrections do not modify this flow equation, as non-analytic terms such as $q_z^2/q^2 \sim \cos^2\theta$ or $q_x^2q_y^2/q^4 \sim \sin^4\theta\sin^2 2\phi$ cannot be generated by the elimination of high energy modes. Therefore it follows that

$$h_i(l) = h_i e^{2l}. \quad (3.129)$$

A second set of flow equations for $\tilde{r}_\varphi$ and u_φ reads

$$\frac{d\tilde{r}_\varphi}{dl} = 2\tilde{r}_\varphi + 3u_\varphi \Lambda^{d-2} A(h_1, h_2) - 3u_\varphi r \Lambda^{d-4} B(h_1, h_2),$$
$$\frac{du_\varphi}{dl} = (4-d)\, u_\varphi - 9u_\varphi^2 \Lambda^{d-4} B(h_1, h_2), \tag{3.130}$$

with

$$A(h_1, h_2) = K_{d-1} \int \frac{\sin^{d-2}\theta d\theta d\phi}{(2\pi)^2} f(\theta, \phi),$$
$$B(h_1, h_2) = K_{d-1} \int \frac{\sin^{d-2}\theta d\theta d\phi}{(2\pi)^2} f^2(\theta, \phi, . \tag{3.131}$$

where $\int \cdots = \int_0^\pi d\theta \int_0^{2\pi} d\phi \cdots$ and

$$f(\theta, \phi) = \frac{1}{1 + h_1 \cos^2\theta + h_2 \sin^4\theta \sin^2(2\phi)}. \tag{3.132}$$

The d-dimensional kernel is given by $K_d = \frac{2\pi^{d/2}}{(2\pi)^d} / \Gamma\left(\frac{d}{2}\right)$. For $h_i \gg 1$ the biggest contribution to the integrals in (3.131) arises from the regions in the vicinity of $\theta \approx \pi/2$ and $\phi \approx 0$ and we find that

$$A(h_1, h_2) \sim \frac{1}{2} \frac{K_{d-1}}{(2\pi)^2} (h_1 h_2)^{-1/2},$$
$$B(h_1, h_2) \sim \frac{1}{4} \frac{K_{d-1}}{(2\pi)^2} (h_1 h_2)^{-1/2}.$$

We introduce the effective coupling constant

$$G = \frac{u_\varphi}{\sqrt{h_1 h_2}}, \tag{3.133}$$

and it holds that

$$\frac{dG}{dl} = (2-d)\, G - 9G^2 \Lambda^{d-4} \frac{K_{d-1}}{4(2\pi)^2}. \tag{3.134}$$

which flows to zero for $d > 2$. Therefore, the mean field behaviour occurs above $d = 2$ [29, 46, 80].

3.4.4 *Estimate of the Mean-Field Regime of Nematic Degrees of Freedom*

Experiments suggest a rather large temperature window in which the mean-field behavior of the nematic degrees of freedom can be seen, as observed from the

Curie-Weiss temperature dependence of the nematic susceptibility. Next, we estimate this temperature range in which the nematic-degrees of freedom are mean-field like.

The scaling $A, B \simeq (h_1 h_2)^{-1/2}$ ceases to be valid when one of the h_i becomes of order 1, see (3.131). We define the length $l = l^*$, such that $h_1(l^*) \approx 1$. Then it follows that $h_2(l^*) = \frac{h_2(0)}{h_1(0)}$, and that for $d = 2$, $u_\varphi(l^*) = \frac{u_\varphi(0)}{h_1(0)}$. The nematic correlation length is $\xi_\varphi(l^*) = \sqrt{h_1(0)}\xi_\varphi(0)$. Our analysis will break down when the nematic correlation length reaches the size of the lattice constant a, i.e. for $\xi_\varphi(l^*) < a$. The mean-field behavior is therefore only valid if

$$\xi_\varphi(0) > \frac{a\Lambda c_s^0}{\sqrt{\lambda_{\text{el}}^2 \mu_1}} \approx \sqrt{\frac{c_s^0}{\lambda_{\text{el}}^2}}, \tag{3.135}$$

where we defined $c_s^0 = c_{11} - c_{12}$. In (3.135) we substituted $\mu_i \sim c_s^0$, see [100] for experimental data, and $a\Lambda \sim 1$. The nematic susceptibility is given by $\chi_{\text{nem}} = r_\varphi^{-1} = \xi_\varphi^2$. Therefore, the condition for the mean-field behavior translates into the following condition for the nematic susceptibility

$$\chi_{\text{nem}} > \frac{c_s^0}{\lambda_{\text{el}}^2}. \tag{3.136}$$

In the next Section we will further discuss the implications of this result.

3.5 Physical Observables

3.5.1 Softening of the Elastic Modulus in the Vicinity of the Nematic Transition

This Section is partially based on the results of the following references [40, 43].

The coupling between the strain and the nematic order parameter has important consequences for the behavior of the elastic constants. To illustrate them, we first consider a simplified Ginzburg–Landau free-energy expansion for the nematic (φ) and orthorhombic ($\varepsilon_s = \partial_x u_x - \partial_y u_y$) degrees of freedom:

$$F = \frac{a}{2}\varphi^2 + \frac{u}{4}\varphi^4 + \frac{c_s^0}{2}\varepsilon_s^2 - \lambda_{\text{el}}\varepsilon_s\varphi - \sigma\varepsilon_s. \tag{3.137}$$

Here, $a \propto T - T_{\text{nem},0}$, with the mean-field nematic transition temperature $T_{\text{nem},0}$, $u > 0$, c_s^0 is the bare shear modulus, λ_{el} is the coupling constant, and σ is the stress, which is the conjugate field to the strain ε_s. The assumption behind this free energy is that the lattice does not soften on its own, i.e. c_s^0 remains finite at all temperatures.

In this case, a harmonic approximation for the lattice degrees of freedom suffices. For a system in which stress is not applied, minimization of the free energy with respect to ε_s gives:

$$\varepsilon_s = \frac{\lambda_{\mathrm{el}}}{c_{s,0}}\varphi, \tag{3.138}$$

implying that nematic order induces lattice distortion. Substituting it back in (3.137) gives:

$$F = \frac{\chi_{\mathrm{nem}}^{-1}}{2}\varphi^2 + \frac{u}{4}\varphi^4 \tag{3.139}$$

where we defined the nematic susceptibility (renormalized by the lattice):

$$\chi_{\mathrm{nem}}^{-1} = a - \frac{\lambda_{\mathrm{el}}^2}{c_s^0} \tag{3.140}$$

While nematic order induces a lattice distortion, nematic fluctuations soften the lattice and affect the shear modulus. Using the thermodynamic definition, the renormalized shear modulus c_s is given by:

$$c_s^{-1} = \frac{\partial \varepsilon_s}{\partial \sigma} \tag{3.141}$$

In the presence of stress, minimization of the free energy gives two coupled equations:

$$\begin{aligned} a\varphi + u\varphi^3 &= \lambda_{\mathrm{el}}\varepsilon_s \\ c_s^0\varepsilon_s &= \lambda_{\mathrm{el}}\varphi + \sigma \end{aligned} \tag{3.142}$$

Taking the derivative with respect to σ yields:

$$\begin{aligned} \left(\frac{\partial\varphi}{\partial\sigma}\right)\left(a + 3u\varphi^2\right) &= \lambda_{\mathrm{el}}\left(\frac{\partial\varepsilon_s}{\partial\sigma}\right) \\ c_s^0\left(\frac{\partial\varepsilon_s}{\partial\sigma}\right) &= \lambda_{\mathrm{el}}\left(\frac{\partial\varphi}{\partial\sigma}\right) + 1 \end{aligned}$$

In the tetragonal phase, $\varphi = 0$. Solving for $\frac{\partial\varepsilon_s}{\partial\sigma}$ and using (3.141) and (3.140) gives the renormalized shear modulus:

$$\begin{aligned} c_s^{-1} &= \left(c_s^0\right)^{-1} + \frac{\lambda_{\mathrm{el}}^2}{\left(c_s^0\right)^2}\chi_{\mathrm{nem}}, \\ \left(\frac{c_s}{c_s^0}\right)^{-1} &= 1 + \frac{\lambda_{\mathrm{el}}^2}{c_s^0}\chi_{\mathrm{nem}}. \end{aligned} \tag{3.143}$$

As expected, $c_s \to 0$ as $\chi_{\text{nem}} \to \infty$, i.e. the nematic and structural transitions are simultaneous. More interestingly, this equation reveals that the shear modulus can be used to probe nematic fluctuations. Measurements of the shear modulus have been performed in a variety of iron-based compounds using resonant ultra-sound or a three-point bending technique [11, 13, 40, 113], revealing a wide regime of strong nematic fluctuations.

In particular, since the nematic transition is mean-field like due to the coupling to the lattice, the nematic susceptibility has the simple Curie-Weiss form $\chi_{\text{nem}} \propto 1/(T - T_S)$, yielding:

$$c_s = c_s^0 \left(1 + \frac{\kappa T_s}{T - T_s}\right)^{-1} = c_s^0 \left(\frac{T - T_s}{T - \theta}\right) \tag{3.144}$$

where κ is a positive dimensionless constant and $\theta = T_s(1 - \kappa) < T_s$. This form agrees well with the experimental observations [11, 40, 113].

Note that the regime in which the mean-field behaviour of the nematic degrees of freedom occurs, derived in (3.136), coincides with the regime of strong elastic modulus renormalization, given by (3.143). From (3.143) we see that when the nematic susceptibility becomes of the order $\chi_{\text{nem}} \sim \frac{c_s^0}{\lambda_{\text{el}}^2}$ strong renormalisation of the elastic modulus will take place. Therefore, we have explicitly shown that the mean field behavior, (3.136) occurs in the entire regime where the suppression of the elastic modulus takes place, see Fig. 3.6 for details. This is exactly what has been observed

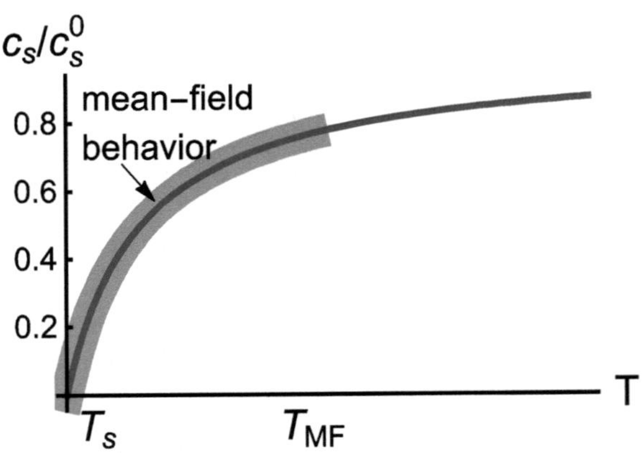

Fig. 3.6 Softening of the shear modulus as one approaches the structural transition and the mean-field regime of the nematic degrees of freedom. The softening follows (3.144); c_s is the physical shear modulus, that can be measured in ultra-sound experiments, while c_s^0 is its high-temperature limit. The regime in which the mean-field behaviour of the nematic degrees of freedom can be expected is shaded *light blue* region. T_{MF} indicates the characteristic temperature where mean field behavior of nematic degrees of freedom sets in upon approaching the transition at T_S. The mean-field regime coincides with the regime of strong elastic modulus renormalization (see the main text for the explanation) Figure from [64]

experimentally [10, 11, 113]. We add that this conclusion is valid regardless of the detailed microscopic origin of nematicity. Curie-Weiss behavior due to the coupling to the lattice is expected in the entire temperature regime where a softening of the elastic constant is observed. For several iron based systems, this regime can be as high as 300–350 K [10, 11, 40, 113].

While we used a simplified Ginzburg–Landau expansion to derive the shear modulus softening, (3.143), we note that this expression is much more general and is valid for an arbitrary form of the nematic action functional $S[\varphi]$, as long as the lattice free energy is harmonic (for the derivation, see [40, 45]). This allows us to make a more quantitative comparison between the Ising-nematic theory and the experimental results. In particular, according to (3.82), the Ising-nematic susceptibility depends on the magnetic fluctuations according to $\int_k \chi_k^2$, where χ_k is the magnetic susceptibility. For a finite temperature nematic transition, the integration is only over momentum $\mathbf{k}$, as the Matsubara sum is dominated by the $n = 0$ contribution, i.e. $\int_k \chi_k^2 \approx T_s \sum_{\mathbf{k}} \chi^2(\mathbf{k})$. Now, this quantity can be probed by NMR via the spin-lattice relaxation rate $1/T_1T$:

$$\frac{1}{T_1T} = \gamma_g^2 \lim_{\omega\to 0} \sum_{\mathbf{k}} F^2(\mathbf{k}) \frac{\mathrm{Im}\chi(\mathbf{k},\omega)}{\omega} \tag{3.145}$$

where γ_g is the gyromagnetic ratio and $F(\mathbf{k})$ is the structure factor of the hyperfine interaction. For a field applied parallel to the FeAs plane, $F(\mathbf{k})$ is peaked at the magnetic ordering vectors $\mathbf{Q}_X = (\pi, 0)$ and $\mathbf{Q}_Y = (0, \pi)$. Furthermore, because of Landau damping, $\chi^{-1}(\mathbf{k},\omega) = \chi^{-1}(\mathbf{k}) - i\omega\Gamma$, implying that $\lim_{\omega\to 0} \frac{\mathrm{Im}\chi(\mathbf{k},\omega)}{\omega} \propto \chi^2(\mathbf{k})$, i.e. $1/T_1T \propto \sum_{\mathbf{k}} \chi^2(\mathbf{k})$. Combining this with (3.82) and (3.143) then yields a scaling relationship between the shear modulus c_s and $1/T_1T$ [43]:

$$\frac{c_s}{c_s^0} = \frac{1}{1 + [a(T_1T) - b]^{-1}} \tag{3.146}$$

Such a scaling behavior has indeed been observed for both Co-doped and K-doped $BaFe_2As_2$ [13, 43], lending strong support for the magnetic origin of the nematic transition in these materials.

3.5.2 *Resistivity Anisotropy*

This Section is partially based on the results of [41].

One of the main experimental probes to assess the properties of nematic state is the in-plane resistivity anisotropy, $\rho_{\mathrm{anis}} \equiv \rho_b - \rho_a$, where b refers to the short in-plane axis, and a refers to the long in-plane axis [7, 22, 76, 102]. Due to the presence of twin domains in the nematic phase, one has to detwin the sample to obtain a single domain and measure a non-vanishing ρ_{anis}. For tensile stress applied along the x-axis, a becomes parallel to x and b becomes parallel to y. In this case, the magnetic ground state at low temperatures corresponds to the $\mathbf{Q}_X = (\pi, 0)$ stripe state. We will assume this to be the case in this section.

From general symmetry arguments, $\rho_{\text{anis}} = \alpha\varphi$, where φ is the nematic order parameter and α is a proportionality constant. The microscopic nature of the nematic phase is encoded not in the fact that $\rho_{\text{anis}} \propto \varphi$, but in the doping and temperature dependence of the proportionality constant α. In the Ising-nematic scenario, the resistivity anisotropy – and thus α – arises from the scattering of the electrons off anisotropic magnetic fluctuations, since fluctuations around $\mathbf{Q}_X = (\pi, 0)$ and $\mathbf{Q}_Y = (0, \pi)$ become different below T_S (see also Sect. 3.5.5). We note that Ising-nematic order also triggers orbital order, which distorts the Fermi surface and also causes a resistivity anisotropy. However, because the sign of this resistivity anisotropy does not agree with the experimentally observed one, we will focus on the anisotropy induced by scattering only.

Following the approach discussed in details in [41], we can calculate ρ_{anis} by solving the semi-classical Boltzmann equation. The quantity obtained from the Boltzmann equation is the non-equilibrium electronic distribution function f_k which, in linear response, can be expanded as $f_k = f_k^0 - \left(\frac{\partial f_k^0}{\partial \varepsilon_k}\right) \Phi_k$, where f_k^0 is the usual Fermi-Dirac equilibrium distribution. Here, the subscript $k = (\mathbf{k}, \lambda)$ involves both momentum $\mathbf{k}$ and band index λ. Upon calculating Φ_k, the conductivity tensor is obtained via $\sigma^{\mu\nu} = -e \sum_k \left(\frac{\partial f_k^0}{\partial \varepsilon_k}\right) v_k^\mu \Phi_k^\nu$, where $\Phi_k \equiv \frac{\Phi_k^\mu E^\mu}{E}$, with μ corresponding to the Cartesian coordinates and $\mathbf{E}$ to the applied electric field. The Boltzmann equation is an integral equation given by [60, 95]:

$$-\frac{e\mathbf{v}_k \cdot \mathbf{E}}{E} \left(\frac{\partial f_k^0}{\partial \varepsilon_k}\right) = \frac{1}{T} \sum_{k'} \left(\Phi_k - \Phi_{k'}\right) f_{k'}^0 \left(1 - f_k^0\right) \tag{3.147}$$
$$\times \left[A_{\text{imp}}^2 \delta\left(\varepsilon_k - \varepsilon_{k'}\right) + A_{\text{spin}}^2 n_B\left(\varepsilon_k - \varepsilon_{k'}\right) \operatorname{Im}\chi_i\left(\mathbf{k} - \mathbf{k}', \varepsilon_k - \varepsilon_{k'}\right)\right]$$

where $n_B(\omega)$ is the Bose-Einstein distribution function and $\chi_i(\mathbf{k}, \omega)$ is the magnetic susceptibility peaked at $\mathbf{Q}_X = (\pi, 0)$ and $\mathbf{Q}_Y = (0, \pi)$. The left side of the Boltzmann equation corresponds to the driving term (due to the application of the electric field) and the right side to the collision term. The latter involves scattering by both impurities (amplitude A_{imp}, here assumed to be the same for all bands) and spin fluctuations (amplitude A_{spin}).

In the limit of large residual resistivity, which seems to be the case of Co-doped $BaFe_2As_2$, the Boltzmann equation can be solved analytically. In this case, the resistivity anisotropy is dominated by the hot spots associated with the magnetic ordering vector selected by the nematic order parameter, i.e. the points of the Fermi surface connected by the magnetic ordering vector $\mathbf{Q}_X = (\pi, 0)$. Using the simplified three band model of (3.60), the four pairs of hot spots are given by the condition

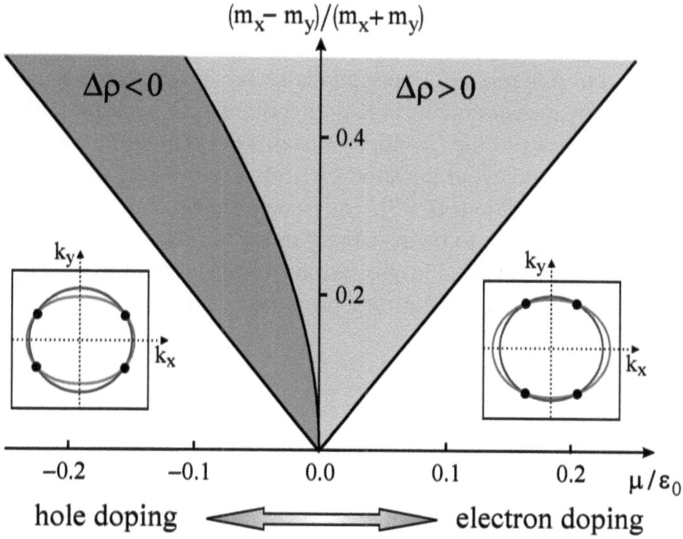

Fig. 3.7 Sign of the resistivity anisotropy $\rho_{\text{anis}} \equiv \Delta\rho = \rho_b - \rho_a$, according to (3.148), as function of the band structure parameters μ (proportional to doping) and $m_x - m_y$ (ellipticity of the electron pockets). The insets illustrate the changes in the positions of the hot spots when doping changes. In these insets, the electron pocket (*red*) has been superimposed to the hole pocket (*blue*) to highlight the hot spots (*black dots*) Taken from [41]

$\varepsilon_{\Gamma,\mathbf{k}} = \varepsilon_{X,\mathbf{k}+\mathbf{Q}_X}$. In the low-temperature limit and near a putative magnetic quantum critical point, the resistivity anisotropy assumes the simple expression:

$$\frac{\rho_{\text{anis}}}{\rho_{\text{imp}}} = \kappa T^{3/2}\left[\frac{\delta_\mu\left(2-\delta_\mu\right)^2}{\delta_m} + \left(4-\delta_\mu\right)\delta_m\right] \tag{3.148}$$

Here, κ is a positive constant and ρ_{imp} is the residual resistivity induced by elastic impurity scattering. The two dimensionless parameters δ_μ and δ_m measure the deviation of the band structure from perfect nesting, with $\delta_\mu \approx 2\mu/E_F$ proportional to the chemical potential (and therefore doping) and $\delta_m \approx m\left(m_x - m_y\right)/\left(2\sqrt{m_x m_y}\right)$ proportional to the ellipticity of the electron pockets. Figure 3.7 shows how ρ_{anis} behaves in the $\left(\delta_\mu, \delta_m\right)$ plane. In general, electron doping ($\delta_\mu > 0$) favors $\rho_{\text{anis}} > 0$ wheres hole doping ($\delta_\mu < 0$) favors $\rho_{\text{anis}} < 0$. This can be understood from a geometrical argument based on the solution of the Boltzmann equation: in the electron-doped side of the phase diagram, the hot spots are closer to the y axis (right inset in Fig. 3.7). As a result, their Fermi-velocities have larger y-components, implying that electrons traveling along the y axis are more strongly scattered by spin fluctuations, i.e. $\rho_{\text{anis}} > 0$. On the other hand, in the hole-doped side of the phase diagram, the hot spots move closer to the x axis (left inset in Fig. 3.7). Consequently, the resistivity due to spin-fluctuation scattering becomes larger along the x axis, i.e. $\rho_{\text{anis}} < 0$.

Although (3.148) itself is derived at low-T and in proximity to a quantum critical point, the sign-change of ρ_{anis} upon doping is a more general feature of the model and not restricted to this regime. Experimentally, this prediction was confirmed by resistivity anisotropy measurements in K-doped $BaFe_2As_2$, which found an opposite sign of ρ_{anis} with respect to the Co-doped systems [8]. The presence of disorder in this model is important to avoid the short-circuiting of the hot spots contribution to the resistivity, as discussed in [60, 95]. The issue of how clean the system has to be in order for this contribution to the resistivity anisotropy be suppressed, remains an open question. Recent numerical investigations of this model found that even in the clean limit a sizable resistivity anisotropy promoted by anisotropic spin fluctuations is observed [15].

3.5.3 Raman Spectroscopy in the Tetragonal Phase

This Section is based on the results of [65, 73].

The signatures of the nematic degrees of freedom manifest themselves in the Raman spectroscopy as well [51, 68], as we will show in this Section. In experiment, one measures the scattering cross-section, which is proportional to the Raman response function. Raman response function is a correlation function between electronic densities weighted by a symmetry factor. The symmetry factor can be changed by changing the polarization of the incoming and outgoing light. This allows one to probe the system (and its ground state and excitations) along different symmetry directions, for example $B_{1g} \sim x^2 - y^2$, $B_{2g} \sim xy$ and A_{1g}. Intuitively, one would therefore expect to see the effect of the nematic phase (characterized by breaking of the symmetry $C_4 \to C_2$, where x and y directions become inequivalent) in the Raman response function. Since the Raman function measures symmetry weighted electron density correlations, and since electrons interact with spin fluctuations which are strong in the vicinity of the transition, the latter will manifest themselves in the Raman response function as corrections to the electron self energy and the Raman vertex, formally expressed in terms of Aslamazov-Larkin diagrams [17].

Raman spectroscopy as a mean of detecting the signatures of nematic degrees of freedom in iron based superconductors was already studied in [52, 72, 106, 115].

Starting from the model of spin-driven nematic phase, described in Sect. 3.3.2, in this Section we will demonstrate, that (i) there is no enhancement of the electronic Raman response function in the B_{2g} channel upon lowering the temperature and approaching the nematic transition from the tetragonal phase, (ii) that the Raman response function develops a pronounced peak at the structural phase transition in the B_{1g} symmetry, and (iii) that there is some response in the A_{1g} channel, which weakens as the temperature is lowered towards the structural transition temperature. This is all in excellent agreement with the experiment [73].

We will first calculate the leading order Aslamazov-Larkin (AL) diagrams and show that they support only the A_{1g} and the B_{1g} symmetry, but not the B_{2g} symmetry, which explains the absence of the fluctuation signal in the B_{2g} channel. However, considering only the leading order diagrams cannot account for the observation of the peak in the Raman response function at the structural transition temperature, as seen in the experiments of [52, 73]. Instead a similar increase only at the magnetic phase transition temperature is expected. Therefore, we go beyond the leading order approach and calculate higher order Aslamazov-Larkin diagrams, which take into account collisions between spin fluctuations that become more and more pronounced as one approaches the nematic transition. This is accomplished by inserting a series of quartic paramagnon couplings, mediated by electrons, into the Raman response function. These quartic couplings mimic the structure of the nematic coupling constant g, calculated in Sect. 3.3.2. The inclusion of such processes leads to a peak in the Raman response function at the structural phase transition in the B_{1g} channel. On the other hand, if we re-sum such higher order AL diagrams in the A_{1g} channel, this will lead to the suppression of the Raman response in the aforementioned channel.

First we will begin by introducing some basic notions of Raman spectroscopy.

3.5.3.1 Raman Response Function – Preliminaries

Raman spectroscopy has been used to study a variety of strongly correlated electronic systems [34], as it can give useful information about lattice, spin or electronic degrees of freedom. In particular, it probes the underlying symmetry of the system, along different directions in the momentum space. This notion has been used to study the symmetry of the electronic excitations [33] and fluctuations in cuprates [17–19, 103], and pnictides.

Raman spectroscopy measures the differential photon scattering cross section, which is proportional to the structure factor S:

$$S_q = -\frac{1}{\pi}\left[1 + n(\omega)\right] \mathrm{Im} R_q. \tag{3.149}$$

The structure factor is related to the imaginary part to the Raman response function R through the fluctuation-dissipation theorem [35]. In the above $n(\omega)$ denotes the Bose-Einstein distribution function, and $q = (\mathbf{q}, \omega)$. Raman spectroscopy is an optical probe, and since the momentum of light is negligible compared to the typical lattice momentum $p \sim 1/a$, where a is the lattice constant, one can safely approximate $\mathbf{q} \approx 0$ in (3.149).

In the regime of non-resonant scattering, the Raman response function is a correlation function between "effective charge density" fluctuations $\tilde{\rho}$,

$$R(\omega) = \int_0^{1/T} d\tau e^{-i\omega\tau} \langle \tilde{\rho}(\tau)\tilde{\rho}(0) \rangle. \tag{3.150}$$

The effective density is the electronic density weighted by the symmetry-dependent form factor that can be tuned via the geometry of the incoming and outgoing light polarization:

$$\tilde{\rho}_k = \sum_{i,k',\sigma} \gamma_{\mathbf{k}'} c^\dagger_{i,k+k',\sigma} c_{i,k',\sigma}. \tag{3.151}$$

The function $\gamma_{\mathbf{k}}$ depends on the incident and scattered photon polarization vectors and the curvature of the bands [35].

Next, we show how to explicitly calculate the Raman response function. In order to do this, we couple an external source field conjugate to the weighted densities and introduce the generating functional W_h:

$$\begin{aligned} W_h &= \frac{1}{Z}\int d\Delta_i d\Psi \mathrm{e}^{-S[\Psi,\Delta_i]-\Psi^\dagger \mathcal{V}_h \Psi}, \\ Z &= \int d\Delta_i d\Psi \mathrm{e}^{-S[\Psi,\Delta_i]}. \end{aligned} \tag{3.152}$$

The matrix $\mathcal{V}_h$, in momentum/frequency, spin and band space is given by

$$\mathcal{V}_{h,k_1k_2\sigma\sigma' ij} = h_{k_1-k_2}\gamma_{\mathbf{k}_1}\delta_{\sigma\sigma'}\delta_{ij}, \tag{3.153}$$

where h is the field conjugate to the effective electronic density. We obtain the Raman response function (3.150) by differentiating the generating functional W_h (3.152) with respect to the conjugate field h:

$$R_q = \left.\frac{\delta^2 W_h}{\delta h_q \delta h_{-q}}\right|_{h=0}. \tag{3.154}$$

The generating functional (3.152) can be expressed as

$$\begin{aligned} W_h &= \frac{1}{Z}\int d\Delta_i d\Psi \mathrm{e}^{\int \Psi^\dagger \mathcal{G}^{-1}_{\Delta,h}\Psi - \frac{1}{2u_s}\int_x (\Delta_X^2+\Delta_Y^2)} \\ \mathcal{G}^{-1}_{\Delta,h} &= \mathcal{G}_0^{-1} - \mathcal{V}_\Delta - \mathcal{V}_h. \end{aligned} \tag{3.155}$$

Because W_h is quadratic in fermionic fields, we integrate out the fermions to obtain:

$$\begin{aligned} W_h &= \frac{1}{Z}\int d\Delta_i \mathrm{e}^{-S_h[\Delta_i]}, \\ S_h[\Delta_i] &= \frac{2}{u_s}\int_x \left(\boldsymbol{\Delta}_X^2 + \boldsymbol{\Delta}_Y^2\right) - \mathrm{Tr}\ln\left(\mathcal{G}^{-1}_{\Delta,h}\right). \end{aligned} \tag{3.156}$$

We expand:

$$\mathrm{Tr}\ln\left(\mathcal{G}^{-1}_{\Delta,h}\right) = \mathrm{Tr}\ln\left(\mathcal{G}^{-1}_{\Delta}\right) - \sum_{n=1}^{\infty}\frac{\mathrm{Tr}\left(\mathcal{G}_\Delta \mathcal{V}_h\right)^n}{n}. \tag{3.157}$$

Using (3.156) and (3.154),

$$R_q = \frac{1}{Z}\int d\Delta_i e^{-S_{\mathrm{eff}}[\Delta_X,\Delta_Y]}\frac{\delta^2}{\delta h_q \delta h_{-q}}\left\{\exp\left[-\mathrm{Tr}\,(\mathcal{G}_\Delta \mathcal{V}_h) - \frac{1}{2}\mathrm{Tr}\,(\mathcal{G}_\Delta \mathcal{V}_h)^2\right]\right\}\Big|_{h=0}. \tag{3.158}$$

Here $S_{\mathrm{eff}}[\Delta_X, \Delta_Y] = S_h[\Delta_i]|_{h=0}$ is the effective action from (3.77). Let us define the matrix

$$\Gamma^q = \frac{\delta \mathcal{V}_h}{\delta h_q}. \tag{3.159}$$

3.5.3.2 Self-energy and Vertex Correction Diagrams

The leading order contributions to the Raman response function can be expressed in terms of the self-energy and vertex correction diagrams [19] (vertex correction diagrams are also known as Maki-Thompson diagrams) which are show in Fig. 3.8. Both of these terms can be obtained by differentiating the second term in the exponential (3.158) twice with respect to h

$$R_q^{V,S} = \frac{1}{Z}\int d\Delta_i e^{-S_{\mathrm{eff}}[\Delta_X,\Delta_Y]}\mathrm{Tr}\left[(\mathcal{G}_\Delta \Gamma)^2\right], \tag{3.160}$$

and by replacing $S_{\mathrm{eff}} \to S_0$, where S_0 is the quadratic part of the action

$$S_0[\Delta_i] = \frac{2}{u_s}\int (\Delta_X^2 + \Delta_Y^2) + \frac{1}{2}\mathrm{Tr}\,(\mathcal{G}_0 \mathcal{V}_\Delta)^2. \tag{3.161}$$

The vertex correction diagram (Maki-Thompson diagram), can be obtained by replacing both $\mathcal{G}_\Delta$ in (3.160) by $\mathcal{G}_\Delta \to \mathcal{G}_0 \mathcal{V}_\Delta \mathcal{G}_0$, see expansion of (3.70):

$$R_q^{V} = \frac{1}{Z}\int d\Delta_i e^{-S_0[\Delta_X,\Delta_Y]}\mathrm{Tr}\left[(\mathcal{G}_0 \mathcal{V}_\Delta \mathcal{G}_0 \Gamma)^2\right]. \tag{3.162}$$

For the self-energy contribution, we replace one of $\mathcal{G}_\Delta$ in (3.160) by $\mathcal{G}_\Delta \to \mathcal{G}_0$, and the other one by $\mathcal{G}_\Delta \to (\mathcal{G}_0 \mathcal{V}_\Delta)^2 \mathcal{G}_0$, such that

$$R_q^{S} = \frac{2}{Z}\int d\Delta_i e^{-S_0[\Delta_X,\Delta_Y]}\mathrm{Tr}\left[\mathcal{G}_0 \Gamma\,(\mathcal{G}_0 \mathcal{V}_\Delta)^2 \mathcal{G}_0 \Gamma\right]. \tag{3.163}$$

We note that the square of the $\gamma_{\mathbf{k}}$ factor appears inside the momentum integrals. Because of this, the self-energy and vertex corrections occur in all symmetry channels. The sum $R^S + R^V$ can be evaluated explicitly in the hot-spot approximation. We found that there were partial cancellation in the A_{1g} and in the B_{1g} channels, and no cancellations in the B_{2g} channel, due to contributions from various hot-spots.

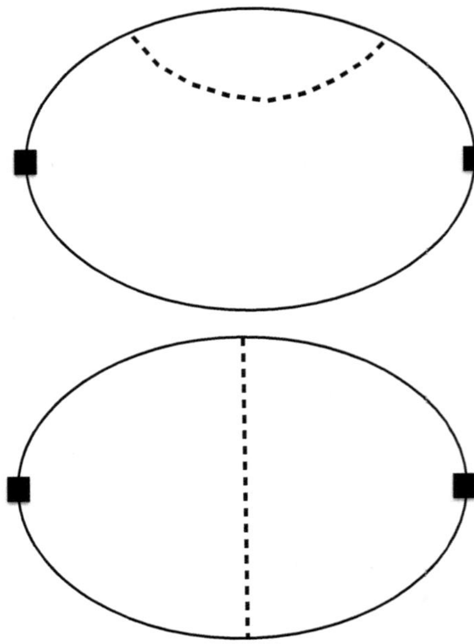

Fig. 3.8 Top: Contribution to the Raman response function that contains the self-energy correction to the fermionic propagator. *Bottom* A diagram that contains a vertex renormalization correction – the so called Maki-Thompson diagram Figure adapted from [65]

In $d = 2$ the calculation leads to

$$R^S + R^V \propto \int_{\mathbf{q}} \frac{1}{r_0 + q^2} \propto \log \xi, \tag{3.164}$$

where $r_0 = \xi^{-2}$, and ξ represents the magnetic correlation length.

To summarize, we have shown that the self-energy and vertex-correction diagrams exist in all symmetry channels and are therefore symmetry insensitive.

3.5.3.3 Leading Order Aslamazov-Larkin Diagrams

In the previous Section we have found that the self-energy and vertex-correction diagrams are not symmetry sensitive, and therefore their consideration alone cannot explain the fact that an anomalous peak in the B_{1g} channel only, and not in the B_{2g} channel was experimentally measured [73]. Therefore, we go beyond and consider Aslamazov-Larkin contributions to the Raman response function [17, 108]. As we will show, these diagrams are symmetry sensitive, but unfortunately they cannot explain the peak of the Raman response function at T_s, in the B_{1g} channel; instead they predict a similar increase at T_N. Leading order Aslamazov-Larkin diagrams arise from differentiating the first term inside the exponential function in (3.158) twice, and from replacing $\mathscr{G}_\Delta \to (\mathscr{G}_0 \mathscr{V}_\Delta)^2 \mathscr{G}_0$, which comes from the expansion of (3.70):

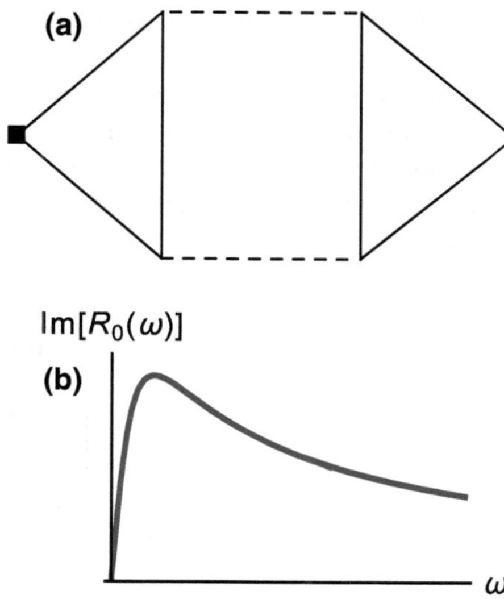

Fig. 3.9 Leading order Aslamazov-Larkin diagram for $d = 2$. **a** leading order diagram, Raman vertices are denoted by *black squares*. **b** plot of the imaginary part of the Raman response function as a function of frequency $\mathrm{Im}R_0(\omega)$, in $d = 2$

$$R_q = \frac{1}{Z}\int d\Delta_i e^{-S_{\text{eff}}[\Delta_X,\Delta_Y]}\left[\mathrm{Tr}\left((\mathscr{G}_0\mathscr{V}_\Delta)^2\mathscr{G}_0\Gamma\right)\right]^2 \tag{3.165}$$

In the above $S_{\text{eff}}[\Delta_X, \Delta_Y] = S_h[\Delta_i]|_{h=0}$ is the effective action from (3.73).

For the time being, we approximate the effective action in (3.165) by a quadratic action (3.161). However, as we will see later on, this assumption, which effectively neglects the interactions between spin fluctuations, is not justified in the theory of spin-driven nematicity.

For multi-band materials, the leading order Aslamazov-Larkin diagram, shown in Fig. 3.9 is given by

$$R_0(\omega) = T\sum_{i=X,Y,n}\int_{\mathbf{q}} \Lambda_i^2(\mathbf{q},\Omega_n,\omega)\chi(\mathbf{q},\Omega_n)\chi(\mathbf{q},\Omega_n-\omega) \tag{3.166}$$

where

$$\begin{aligned}
\Lambda_i(\mathbf{q},\Omega,\omega) &= \Lambda_i^{(1)}(\mathbf{q},\Omega,\omega) + \Lambda_i^{(2)}(-\mathbf{q},-\Omega,-\omega),\\
\Lambda_i^{(1)}(\mathbf{q},\Omega,\omega) &= T\sum_n\int_{\mathbf{k}} \gamma_{\mathbf{k}} G_\Gamma(\mathbf{k},\nu_n-\omega)G_\Gamma(\mathbf{k},\nu_n)\times G_i(\mathbf{k}-\mathbf{q},\nu_n-\Omega),\\
\Lambda_i^{(2)}(\mathbf{q},\Omega,\omega) &= T\sum_n\int_{\mathbf{k}} \gamma_{\mathbf{k}} G_i(\mathbf{k},\nu_n-\omega)G_i(\mathbf{k},\nu_n)\times G_\Gamma(\mathbf{k}-\mathbf{q},\nu_n-\Omega),
\end{aligned} \tag{3.167}$$

is the triangle vertex, similar to [93].

Raman Response in Different Symmetries

We will now show that the leading order Aslamazov-Larkin diagram supports only the A_{1g} and the B_{1g} channels, and is therefore symmetry sensitive. Higher order corrections to the leading order Aslamazov-Larkin diagram that will be discussed later do not alter this symmetry-based analysis. The symmetry factors γ in various channels are

$$\gamma_{\mathbf{k}}^{A_{1g}} = \frac{1}{2}\left\{\frac{\partial^2\epsilon(\mathbf{k})}{\partial k_x^2} + \frac{\partial^2\epsilon(\mathbf{k})}{\partial k_y^2}\right\},$$
$$\gamma_{\mathbf{k}}^{B_{1g}} = \frac{1}{2}\left\{\frac{\partial^2\epsilon(\mathbf{k})}{\partial k_x^2} - \frac{\partial^2\epsilon(\mathbf{k})}{\partial k_y^2}\right\},$$
$$\gamma_{\mathbf{k}}^{B_{2g}} = \frac{\partial^2\epsilon(\mathbf{k})}{\partial k_x \partial k_y}, \tag{3.168}$$

where $\epsilon(\mathbf{k})$ denotes the band dispersion.

We analyze the Aslamazov-Larkin diagram, given by (3.166) and (3.167). Let us examine the symmetry of the terms in (3.166) which arise from

$$R_0^{(11)}(\omega) = T \sum_{i=X,Y,n} \int_{\mathbf{q}} \left[\Lambda_i^{(1)}(\mathbf{q}, \Omega_n, \omega)\right]^2 \chi(\mathbf{q}, \Omega_n)\chi(\mathbf{q}, \Omega_n - \omega). \tag{3.169}$$

We rewrite the term (3.169) in the following form

$$R_0^{(11)}(\omega) = \frac{T}{2}\sum_n \int_{\mathbf{q}}\int_{\mathbf{k}}\int_{\mathbf{p}} \gamma_{\mathbf{k}}\gamma_{\mathbf{p}}\chi(\mathbf{q}, \Omega_n)\chi(\mathbf{q}, \Omega_n - \omega)$$
$$\times(E_{A_{1g}}(\omega, \Omega_n, \mathbf{k}, \mathbf{q})E_{A_{1g}}(\omega, \Omega_n, \mathbf{p}, \mathbf{q})$$
$$+E_{B_{1g}}(\omega, \Omega_n, \mathbf{k}, \mathbf{q})E_{B_{1g}}(\omega, \Omega_n, \mathbf{p}, \mathbf{q})), \tag{3.170}$$

where we have combined the appropriate combinations of Green's functions according to their symmetry class into

$$E_{A_{1g}}(\omega, \Omega_n, \mathbf{k}, \mathbf{q}) = T\sum_m G_\Gamma(\mathbf{k}, \nu_m - \omega)G_\Gamma(\mathbf{k}, \nu_m)G^{(+)}(\mathbf{k}-\mathbf{q}, \nu_m - \Omega_n),$$
$$E_{B_{1g}}(\omega, \Omega_n, \mathbf{k}, \mathbf{q}) = T\sum_m G_\Gamma(\mathbf{k}, \nu_m - \omega)G_\Gamma(\mathbf{k}, \nu_m)G^{(-)}(\mathbf{k}-\mathbf{q}, \nu_m - \Omega_n), \tag{3.171}$$

and where $G^{(\pm)} = G_X \pm G_Y$. It is easy to verify that the functions in (3.171) posses the desired symmetries. For example, B_{1g} symmetry changes sign under $\pi/2$-rotation, but not under the reflection with respect to one of the axes, while the

A_{1g} does not change under either of the operations. From (3.171), we clearly notice that the response will be non-zero only in the A_{1g} or the B_{1g} symmetry channels. One can show that all other terms in (3.166) support the A_{1g} or the B_{1g} symmetry only. Leading order Aslamazov-Larkin diagrams therefore do not support the B_{2g} symmetry channel. Finally we would like to comment that this symmetry sensitivity is robust, and holds irrespective to weather or not we used the hot-spot approximation to explicitly evaluate the diagrams.

Calculation of the Leading Order Aslamazov-Larkin Diagram

We follow the calculation of [17]. After the analytic continuation to real frequencies, we find that

$$\mathrm{Im} R_0(\omega + i0^+) = \int_{-\infty}^{\infty} \frac{d\epsilon}{\pi} \left[n(\epsilon) - n(\epsilon+\omega)\right] \times \int_{\mathbf{q}} \mathrm{Im}\left[\chi^R(\epsilon, \mathbf{q})\right] \mathrm{Im}\left[\chi^R(\epsilon+\omega, \mathbf{q})\right], \tag{3.172}$$

with the spin propagator given by:

$$\chi^R(\mathbf{q}, \Omega) = \frac{1}{r_0 + \mathbf{q}^2 - i\Omega}. \tag{3.173}$$

Next we perform the momentum integral in (3.172) explicitly in $d = 2$ to get

$$\mathrm{Im}\left[R_0(\omega + i0^+)\right]_{d=2} = \int_0^{\infty} d\epsilon \left[n(\epsilon_+) - n(\epsilon_-)\right] \frac{\epsilon_+ \epsilon_-}{\epsilon_+^2 - \epsilon_-^2} \times \left[F(\epsilon_+) - F(\epsilon_-)\right], \tag{3.174}$$

where

$$F(x) = \frac{1}{x}\left(\arctan\frac{r_0}{x} - \frac{\pi}{2}\mathrm{sgn}(x)\right). \tag{3.175}$$

We have defined $\epsilon_\pm = \epsilon \pm \omega/2$.

The lineshape given by (3.174) is shown in Fig. 3.9. First we consider the high-temperature regime $T \gg r_0$. In this regime $R_0(\omega)_{d=2} \simeq \frac{\omega T}{r_0^2}$ for small frequencies ω, while the maximum of the Raman response function scales as $R_0^{\max}(\omega)_{d=2} \simeq \frac{T}{r_0}$.

In summary, we have shown that the leading order Aslamazov-Larkin diagram does not exist in the B_{2g} channel. On the contrary, it gives a finite response in the A_{1g} and the B_{1g} channels. It predicts the divergence of the Raman response function only at the onset of the magnetic phase, at T_N. We therefore need to go beyond the leading order Aslamazov-Larkin diagram.

3.5.3.4 Higher Order Aslamazov-Larkin Diagrams

In the previous Section, we considered the quadratic action approximation for S_{eff} in (3.165). Now we will go beyond this approximation and include the full quartic action to evaluate the Raman response function [65]. This amounts to inserting a series of fermionic boxes that resemble the structure of the nematic coupling constant g into the leading order Aslamazov-Larkin diagram in the B_{1g} symmetry channel. These fermionic boxes effectively mediate the collisions between spin fluctuations which were not taken into account in the leading order Aslamazov-Larkin diagram. The collisions between spin fluctuations become more and more important as one approaches the transition, as the fluctuations get stronger.

We start from (3.165), and include *quartic* terms in the effective action (rather than only quadratic):

$$R_q = \frac{1}{Z}\int d\Delta_i e^{-S_{\text{eff}}[\Delta_i]}\left[\text{Tr}\left((\mathscr{G}_0\mathscr{V}_\Delta)^2\mathscr{G}_0\Gamma\right)\right]^2 \tag{3.176}$$

where

$$S_{\text{eff}}\left[\boldsymbol{\Delta}_i\right] = S_0\left[\boldsymbol{\Delta}_i\right] + \frac{1}{4}\text{Tr}\left(\mathscr{G}_0\mathscr{V}_\Delta\right)^4, \tag{3.177}$$

and

$$S_0\left[\boldsymbol{\Delta}_i\right] = \frac{1}{2u_s}\int_x\left(\boldsymbol{\Delta}_X^2 + \boldsymbol{\Delta}_Y^2\right) + \frac{1}{2}\text{Tr}\left(\mathscr{G}_0\mathscr{V}_\Delta\right)^2. \tag{3.178}$$

We expand the exponential according to

$$e^{-\frac{1}{4}\text{Tr}(G_0\mathscr{V}_\Delta)^4} \approx \sum_{m=0}^{\infty}\frac{1}{m!}\left[\frac{-1}{4}\text{Tr}\left(\mathscr{G}_0\mathscr{V}_\Delta\right)^4\right]^m \tag{3.179}$$

to get that

$$R_q = \sum_{m=0}^{\infty}\frac{1}{m!}R_q^{(m)}, \tag{3.180}$$

where the following terms were averaged with respect to the Gaussian collective spin action:

$$R_q^{(m)} = \left\langle\left[\frac{-1}{4}\text{Tr}\left(\mathscr{G}_0\mathscr{V}_\Delta\right)^4\right]^m\left[\text{Tr}\left((\mathscr{G}_0\mathscr{V}_\Delta)^2\mathscr{G}_0\Gamma\right)\right]^2\right\rangle_{S_0}.$$

Next we perform contractions of the Δ fields. This results in a series of diagrams that look like the leading order Aslamazov-Larkin diagram with an arbitrary number of fermionic boxes which have been inserted in between the triangular fermionic loops containing the Raman vertices, depicted in Fig. 3.10.

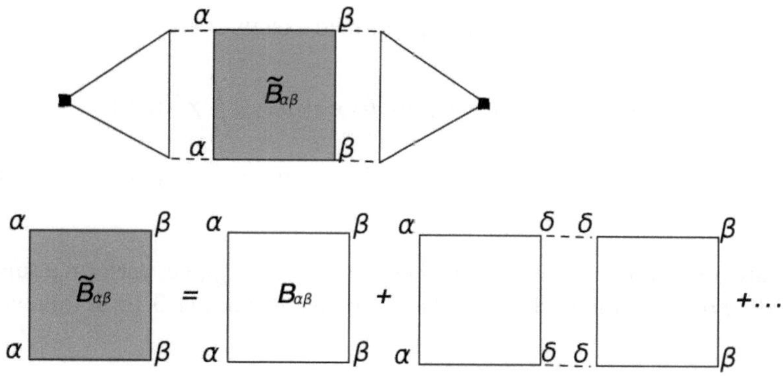

Fig. 3.10 Re-summation of the Raman response function. The re-summed box $\tilde{B}_{\alpha\beta}$ is shaded *light blue*. The first index of the matrix B denotes the type $\alpha = X, Y$ of entering spin fluctuations, and the second index denotes the type of exiting spin fluctuations. The *blue shaded box* is inserted into the leading order AL diagram Figure adapted from [65]

We next show that the re-summation of such boxed Aslamazov-Larkin diagrams leads to the maximum of the Raman response function at the nematic/structural phase transition in the B_{1g} channel.

The fermionic box can be described by two indices $B_{\alpha\beta}$: the first index stands for the type of incoming spin fluctuations, this can be either X or Y, and the second index denotes the type of exiting spin fluctuation.

The summation of all such diagrams leads to:

$$R(\omega) = R_0(\omega) + T^2 \sum_{\Omega,\Omega'} \int_{\mathbf{q},\mathbf{q}'} \Lambda_\alpha(\omega, \Omega, \mathbf{q}) \chi(\mathbf{q}, \Omega) \chi(\mathbf{q}, \Omega - \omega) \tilde{B}_{\alpha\beta}(\mathbf{q}, \mathbf{q}', \Omega, \Omega', \omega) \\ \times \chi(\mathbf{q}', \Omega') \chi(\mathbf{q}', \Omega' - \omega) \Lambda_\beta(\omega, \Omega', \mathbf{q}'). \tag{3.181}$$

We calculate the box $B_{\alpha\beta}$ at momenta $\mathbf{q}, \mathbf{q}' \approx \mathbf{Q}_{X,Y}$ and zero frequencies, which is a good approximation for small incoming Raman frequency ω, and in the classical regime. The Raman response function is then given by

$$R(\omega) \approx R_0(\omega) + \int_{\mathbf{q},\mathbf{q}'} \Lambda_\alpha(\omega, 0, \mathbf{q}) \tilde{B}_{\alpha\beta} \times \chi^2(\mathbf{q}, 0) \chi^2(\mathbf{q}', 0) \Lambda_\beta(\omega, 0, \mathbf{q}'), \tag{3.182}$$

where $R_0(\omega)$ stand for the leading order Aslamazov-Larkin diagram.

The following symmetries of the fermionic triangles hold in the tetragonal phase

$$\Lambda_X^{B_{1g}} = -\Lambda_Y^{B_{1g}}, \\ \Lambda_X^{A_{1g}} = \Lambda_Y^{A_{1g}}. \tag{3.183}$$

The matrix multiplication can then be performed exactly

$$\tilde{R}_{B_{1g}}(\omega) = R_0(\omega) + R_0(\omega)(\tilde{B}_{XX} - \tilde{B}_{XY}) \int_{\mathbf{q}} \chi^2(\mathbf{q}, 0),$$
$$\tilde{R}_{A_{1g}}(\omega) = R_0(\omega) + R_0(\omega)(\tilde{B}_{XX} + \tilde{B}_{XY}) \int_{\mathbf{q}} \chi^2(\mathbf{q}, 0). \tag{3.184}$$

In particular, next we want to calculate the full box $\tilde{B}_{\alpha\beta}$, i.e. perform a sum over the leading box-diagrams within the $1/\mathcal{N}$ expansion, see Fig. 3.10. It follows that

$$\begin{aligned}\tilde{B}_{\alpha\beta} &= B_{\alpha\beta} + B_{\alpha\delta} B_{\delta\beta} \int_{\mathbf{q}'} \chi^2(\mathbf{q}', 0) + \cdots \\ &= \sum_{m=1}^{\infty} (B^m)_{\alpha\beta} \left(\int_{\mathbf{q}} \chi^2(\mathbf{q}, 0) \right)^{m-1}.\end{aligned} \tag{3.185}$$

The matrix B was calculated using (3.180) and by explicitly evaluating the tedious $SU(\mathcal{N})$ trace pre-factors (which arise from various contractions of products of λ matrices, the generators of the $SU(\mathcal{N})$ algebra, in (3.69)), see [65] for details that will not be presented here. It was shown that the matrix B of irreducible boxes is given by

$$B = -\tfrac{\mathcal{N}}{8} \begin{pmatrix} g_{XX} & g_{XY} \\ g_{XY} & g_{XX} \end{pmatrix} \tag{3.186}$$

where we defined

$$\begin{aligned} g_{XX} &= \int_k G_{\Gamma,k}^2 G_{X,k}^2, \\ g_{XY} &= \int_k G_{\Gamma,k}^2 G_{X,k} G_{Y,k}, \end{aligned} \tag{3.187}$$

and exploited the fact that $\int_k G_{\Gamma,k}^2 G_{X,k}^2 = \int_k G_{\Gamma,k}^2 G_{Y,k}^2$.

The mth power of the matrix B is

$$B^m = \frac{1}{2} \left(\frac{-\mathcal{N}}{8} \right)^m \begin{pmatrix} (g_+^m + g_-^m) & (g_+^m - g_-^m) \\ (g_+^m - g_-^m) & (g_+^m + g_-^m) \end{pmatrix}, \tag{3.188}$$

with $g_\pm = g_{XX} \pm g_{XY}$. Therefore

$$\tilde{R}_{B_{1g}}(\omega) = R_0(\omega) \sum_{m=0}^{\infty} \left(\frac{-\mathcal{N} g_-}{8} \right)^m \left(\int_{\mathbf{q}} \chi^2(\mathbf{q}, 0) \right)^m$$

$$= R_0(\omega) + R_0(\omega) \frac{g \int_{\mathbf{q}} \chi^2(\mathbf{q}, 0)}{1 - g \int_{\mathbf{q}} \chi^2(\mathbf{q}, 0)},$$

$$\tilde{R}_{A_{1g}}(\omega) = R_0(\omega) \sum_{m=0}^{\infty} \left(\frac{-\mathcal{N} g_+}{8} \right)^m \left(\int_{\mathbf{q}} \chi^2(\mathbf{q}, 0) \right)^m$$

$$= R_0(\omega) + R_0(\omega) \frac{u \int_{\mathbf{q}} \chi^2(\mathbf{q}, 0)}{1 + u \int_{\mathbf{q}} \chi^2(\mathbf{q}, 0)}, \quad (3.189)$$

where

$$g = -\frac{\mathcal{N}}{16} \int_k G^2_{\Gamma,k} (G_{X,k} - G_{Y,k})^2 \quad (3.190)$$

is precisely the nematic coupling constant of (3.78), and u is the other quartic term in (3.78). In [42] it was found that $u > 0$ for a finite-temperature magnetic transition. Therefore, according to (3.189), the Raman response function in the A_{1g} channel will be suppressed. For the response in the B_{1g} channel, after performing the analytic continuation, we get that

$$\mathrm{Im} \tilde{R}_{B_{1g}}(\omega) = \mathrm{Im} \left[R_0(\omega) \right] \left(1 + g \chi^{\mathrm{el}}_{\mathrm{nem}} \right), \quad (3.191)$$

where

$$\chi^{\mathrm{el}}_{\mathrm{nem}} = \frac{\int_{\mathbf{q}} \chi^2(\mathbf{q}, 0)}{1 - g \int_{\mathbf{q}} \chi^2(\mathbf{q}, 0)} \quad (3.192)$$

is the electronic contribution to the nematic susceptibility in the limit of large N.

We remind ourselves that the susceptibility of the nematic order parameter is given by (3.82):

$$\chi_{\mathrm{nem}} = \frac{\int_q \chi_q^2}{1 - g_{\mathrm{stat}} \int_q \chi_q^2}, \quad (3.193)$$

where $g_{\mathrm{stat}} = g + \frac{\gamma_{\mathrm{el}}^2}{c_s^0}$ is the static limit of the frequency and momentum dependent nematic coupling constant. However, this enhancement of the static nematic coupling constant does not appear in the Raman response function, due to the fact that the Raman response operates in the dynamical limit ($\mathbf{q} = 0$ and finite ω). In [51, 72] it was pointed out that the static and the dynamic limits do not commute.

As was pointed out in [53], the enhancement of the static nematic coupling constant (3.83) does not enter the Raman response, due to the fact that the Raman response operates in the dynamical limit ($\mathbf{q} = 0$ and finite ω), and the static and dynamic limits do not commute [53]. If there were no nemato-elastic coupling, i.e. $\gamma_{\mathrm{el}} = 0$, this would then lead to the divergence of the Raman response function at the structural phase transition, due to the same structure of the denominator as for the nematic susceptibility. Now let us consider finite nemato-elastic coupling $\gamma_{\mathrm{el}} \neq 0$.

The location of the structural/nematic transition is determined by the condition of diverging nematic susceptibility:

$$\left(g + \frac{\gamma_{\text{el}}^2}{c_s^0}\right) \int_q \chi_q^2 = 1. \tag{3.194}$$

Consequently the Raman response function in the B_{1g} channel, given by (3.191), should have a maximum rather than a divergence at the nematic phase transition, in the case where magnetic and structural phase transitions are separated [66, 69, 98].

A similar type of re-summation to that presented above has also been performed in [68] for the B_{1g} channel.

3.5.3.5 Comparison to Experiment: $BaFe_{2-2x}Co_{2x}As_2$

A detailed comparison with experimental observations for $BaFe_{2-2x}Co_{2x}As_2$ was performed in [73] and yields excellent agreement with the presented theory. In particular, in the tetragonal phase $T > T_s$ in the B_{1g} channel the Raman response function displays a maximum at the structural phase transition at T_s (for details see Fig. 3.11). The B_{2g} spectra were measured only at a few temperatures and found to be nearly temperature independent in agreement with the theoretical analysis of Sect. 3.5.3.3 which shows that the fluctuation contribution to the Raman response in the B_{2g} channel should be zero. The analysis of Sect. 3.5.3.4 predicts a finite fluctuation contribution to the Raman response in the A_{1g} channel, although not as strong as in the

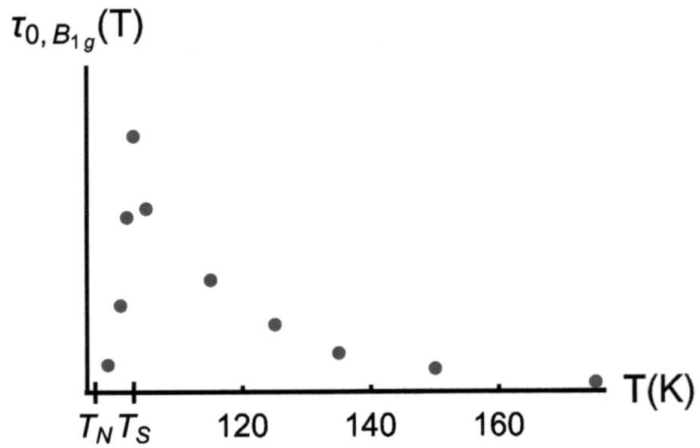

Fig. 3.11 The initial slope $\tau_{0,B_{1g}}(T) = \frac{\partial R_{B_{1g}}(\omega)}{\partial \omega}|_{\omega=0}$ of the Raman response function in the B_{1g} channel plotted as a function of temperature T, for the doping level $x = 0.025$ (for which the structural and magnetic transitions are separated). The maximum of the Raman response function coincides with the structural transition temperature at $T_s \simeq 103K$ as deduced by the appearance of twin boundaries [73]. One should also note that the Raman response function falls of rapidly inside the nematic phase ($T < T_s$), and the signal disappears around the magnetic transition temperature $T_N \simeq 98K$ Experimental data taken from [73]

B_{1g} channel, due to the suppression of the fluctuation contribution coming from the higher-order Aslamazov-Larkin diagrams, (3.189).

The more rapid decrease of the Raman response function inside of the nematic phase, as seen in Fig. 3.11 can be explained as follows. Close to T_s, we expect (3.191) to hold qualitatively also inside the nematic phase, $T_N < T < T_s$. We argue that $\chi^{el}_{nem}(0)$ and, according to (3.191), the Raman amplitude is smaller in the ordered (nematic) phase than in the disordered (tetragonal) phase. This explains the more rapid reduction of the Raman response as one enters the nematic state.

3.5.4 Raman Resonance Mode in the Superconducting State

This Section is based on the results of [53].

Now, we consider the signatures of nematic fluctuations in the Raman response function in the superconducting phase.

The Raman response in the superconducting state is usually characterized by the emergence of a pair-breaking peak at twice the superconducting gap amplitude, an effect that has been used successfully to determine the momentum dependence of the gap [34]. The pair-breaking peak is determined by the form-factor weighted density-density response of otherwise non-interacting Bogoliubov quasi-particles. In our discussion of the Raman response in the normal state, we learned that nematic fluctuations directly impact the spectrum. While we derived the dynamic Raman response from a microscopic theory, we will, in what follows, perform a phenomenological Landau-theory description. In the normal state, the form of $R(\omega)$ in the B_{1g} channel is expected to be of the form

$$R_{B_{1g}}(\omega) = \gamma^2 \frac{m^* \rho_0}{m} \frac{Y(\omega)}{1 + \frac{1}{5} F_2^s Y(\omega)} + R_{\text{inc.}} \tag{3.195}$$

Here, m^*/m and F_2^s are the effective mass ratio and the Landau coefficient in the $l = 2$ charge-Pomeranchuk channel, respectively. The factor $\frac{1}{5}$ originates from the usual convention of Fermi-liquid parameters, where for arbitrary l holds $1 + \frac{1}{2l+1} F_l^{s,a}$ in the denominator. γ and R_{inc} are the vertex that describes the coupling of quasi-particles to photons and the contribution from incoherent, non-quasiparticle excitations, respectively. Both are nontrivial due to the fact that the effective Raman density $\tilde{\rho}_k$, introduced above, is not a conserved quantity [79]. For low frequencies, the dynamic term $Y(\omega) = 1 + i\Gamma\omega$ is expected to follow from the corresponding kinetic equation of $\tilde{\rho}_k$. ρ_0 is the bare $d-$wave weighted electron-pocket density of states $\rho_0 = \frac{1}{N} \sum_{\mathbf{k}} \gamma^2_{B_{1g}}(\mathbf{k}) \delta(\epsilon_F - \epsilon_{\mathbf{k}})$. Equation (3.195) is the usual Fermi liquid formulation for the response in the $l = 2$ charge-Pomeranchuk channel. Near the nematic instability it holds that $F_2^s \to -5$, i.e. the Raman response becomes singular with diverging low-frequency slope.

Fermi liquid theory has been generalized to the superconducting state [79]. Thus, it is interesting to explore the behavior of the Raman response in the superconducting state near a $l = 2$ charge-Pomeranchuk instability. We will see that in this case, strong *nematic correlations* between the quasiparticles transforms the pair-breaking peak into a sharp *nematic resonance* discussed in [53]. It turns out that the energy of this sharp mode, which is below twice the superconducting gap, stays finite even at the nematic quantum critical point and that the spectral weight of the mode is strongly enhanced. These findings are consistent with experiments on electron-doped iron based superconductors [53, 106] and provides direct evidence of nematic correlations in the superconducting phases and reveals that Bogoliubov quasiparticles couple strongly to nematic fluctuations.

3.5.4.1 Nematic Resonance

The emergence of a sharp excitonic peak in the dynamic excitation spectrum of a superconductor is a well known phenomenon in correlated superconductors. We will see that the nematic resonance is rather similar in nature to the spin-resonance, observed in neutron scattering experiments in Fe SC and cuprates superconductors [14, 21, 57, 61, 84, 97]. There is even a close formal analogy between both phenomena, if one follows the interpretation for the neutron mode put forward in [1, 38]. The singular behavior of the imaginary part of a bare response function that is caused by the BCS coherence factors of the superconducting state, is amplified by the vicinity to a quantum critical point.

The generalization of the Fermi-liquid description, (3.195), to the superconducting state is straightforward and given by

$$R_{B_{1g}}(\omega) \equiv \mathrm{Im}\chi_{\mathrm{nem}}(\omega) = \gamma^2 \rho_0 \mathrm{Im}\left(\frac{Y(\omega)}{1+\frac{1}{5}F_2^s Y(\omega)}\right). \tag{3.196}$$

The key difference is the dynamic function $Y(\omega)$ that can be obtained from the behavior of free electrons in a superconducting state. Thus, we follow the usual logic of Fermi liquid theory in the superconducting state that uses Fermi liquid parameters of the normal state [79]. It follows (see [53]):

$$\mathrm{Im}Y(\omega) = \frac{\pi}{2}\frac{(2\Delta)^2\,\theta\left(\omega^2-(2\Delta)^2\right)}{\omega\sqrt{\omega^2-(2\Delta)^2}}. \tag{3.197}$$

evaluated at $T = 0$. We assume a constant superconducting gap Δ on the electron pockets that are expected to dominate the B_{1g}-Raman response. The result for $Y(\omega)$, up to the density of state coefficient the usual Raman response one for a fully gapped superconductor [70]. Thus, it includes the mentioned pair-breaking peak at 2Δ that can be identified as a square-root divergence of $\mathrm{Im}Y(\omega)$ [34]. We also need the real part of $Y(\omega)$ shown in Fig. 3.12a. For $|\omega| < 2\Delta$, which is the relevant part of our consideration, it follows after Kramers–Kronig transformation

$$\mathrm{Re}Y(\omega) = \frac{(2\Delta)^2 \arcsin\left(\frac{\omega}{2\Delta}\right)}{\omega\sqrt{(2\Delta)^2 - \omega^2}}. \tag{3.198}$$

We are now in a position to analyze the Raman response in the superconducting state near the Pomeranchuk instability, i.e. for $\frac{1}{5}F_1^s$ close to -1.

Because of the square-root divergence of $\mathrm{Re}Y(\omega) \to +\infty$ for $\omega \to 2\Delta$ from below, i.e. in the regime where $\mathrm{Im}Y(\omega) = 0$ (see Fig. 3.12a) holds that the denominator in (3.196) is bound to vanish for some frequency $0 < \Omega_r < 2\Delta$, as long as $F_2^s < 0$. Thus for $\omega = \Omega_r$ determined via

$$1 + \frac{1}{5}F_2^s \mathrm{Re}Y(\Omega_r) = 0. \tag{3.199}$$

In the case without impurity scattering and at $T = 0\,K$, the Raman mode is arbitrarily sharp because the particle-hole continuum only starts at 2Δ. Near the resonance the electronic Raman response is:

$$R_{B_{1g}}(\omega) = Z_r \delta(\omega - \Omega_r) \tag{3.200}$$

with spectral weight $Z_r = \gamma^2 \zeta\left(\frac{\Omega_r}{2\Delta}\right)$ determined by $\zeta(x) = \frac{\pi x\left(1-x^2\right)\arcsin(x)}{x\sqrt{1-x^2}+\left(2x^2-1\right)\arcsin(x)}$. As the resonance mode approaches 2Δ its weight vanishes as $Z_r \approx \gamma^2 \pi \frac{\Omega_r - 2\Delta}{\Delta}$ and eventually the total Raman response is dominated by the pair breaking peak. However, for a typical value of $\Omega_r = \frac{3}{4}2\Delta$, the weight is $Z_r \approx 1.45\gamma^2$, which is much larger than the total weight $\pi^2\rho_0\gamma^2\Delta/2$ of the BCS pair-breaking peak because $\rho\Delta \ll 1$. From (3.199) follows that the resonance frequency is determined by the condition

$$\Omega_r = 2\Delta \sin\theta(F_2^s), \tag{3.201}$$

where $\theta(F_2^s)$ obeys the equation $\sin 2\theta = -\frac{2}{5}F_2^s\theta$.

In Fig. 3.12b we show the Raman response for different values of the Landau parameter. We included finite width Γ that may be the result of impurity scattering or due to thermal excitations of quasiparticles. For F_2^s somewhat away from the critical value one finds that the pair-breaking peak keeps its overall line shape yet it increases in its weight. However, as soon as $2\Delta - \Omega_r$ becomes larger than the line width Γ, a sharp resonance peak emerges that is well separated from the continuum above 2Δ.

As discussed in detail in [53] one has to be careful in the analysis of the relevant Landau parameter if one includes a coupling to the elastic degrees of freedom. Suppose there is a purely electronic contribution to the Landau parameter F_2^s. The static nematic response will then be modified according to

$$F_2^s \to F_{2,\mathrm{tot}}^s = F_2^s - \frac{5\rho_0\gamma^2}{c_s^0}, \tag{3.202}$$

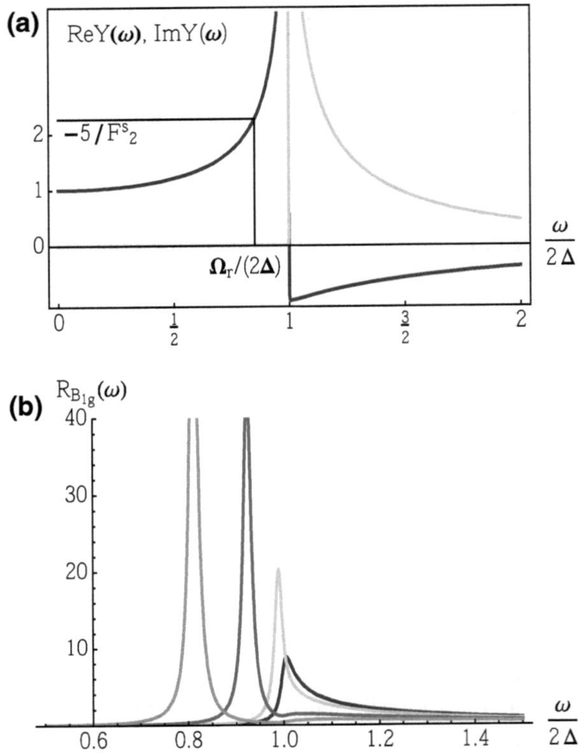

Fig. 3.12 **a** Real and imaginary parts of the dynamic Raman response $Y(\omega)$ for an isotropic gap. The full Raman response (3.196) develops a resonance at $\Omega_r < 2\Delta$ when Re$Y(\Omega_r)$ reaches the threshold $-5/F_2^s$ with Landau parameter F_2^s that is expected to be negative and close to -5 at the nematic instability. **b** Development of the nematic resonance in the superconducting B_{1g} Raman response $R_{B_{1g}}$ for different values of the Landau parameter F_2^s Taken with minor adaption from [53]

pushing the system closer towards a nematic instability. However, as discussed already in (3.83) the Raman response is governed by typical frequencies ω that are larger than $c_s q$ with sound velocity c_s and transferred momentum in the light scattering q. In this regime the Raman response is continued to be governed by the purely electronic contribution F_2^s. Thus, the Raman signature is less singular than the static, thermodynamic response probed via elastic constant measurements. Yet, it allows to draw conclusions for the purely electronic contributions to the nematic fluctuation spectrum.

The key conclusions of this analysis are twofold: (i) the excellent agreement between the theory of a Raman resonance of [53] and Raman experiments in the superconducting state strongly supports the view that nematic fluctuations couple strongly to Bogoliubov quasiparticles, and (ii) the theory only relies on the vicinity to an electronic driven nematic instability, regardless of the detailed microscopic nature. Thus, it can be used as a rather general guiding principle for electronic nematicity, provided one considers a fully gapped superconductor without quasiparticle excitations below 2Δ.

3.5.5 Magnetic Spectrum

This Section is partially based on the results of [42]. The onset of long-range nematic order has important feedback effects on the magnetic spectrum. Indeed, from (3.79), the quadratic part of the magnetic action in the nematic state becomes:

$$S_{\text{eff}} = \int_q \left(\tilde{\chi}_q^{-1} + \varphi\right) \boldsymbol{\Delta}_X^2 + \int_q \left(\tilde{\chi}_q^{-1} - \varphi\right) \boldsymbol{\Delta}_Y^2 \tag{3.203}$$

where $\varphi \neq 0$ is the nematic order parameter and $\tilde{\chi}_q^{-1} = \chi_q^{-1} + i\lambda$ is the magnetic susceptibility renormalized by the Gaussian magnetic fluctuations $i\lambda \propto \left\langle \boldsymbol{\Delta}_X^2 + \boldsymbol{\Delta}_Y^2 \right\rangle$. Thus, an immediate consequence of the onset of nematicity is that the magnetic fluctuation spectrum becomes different around the two magnetic ordering vectors $\mathbf{Q}_X = (\pi, 0)$ and $\mathbf{Q}_Y = (0, \pi)$. For instance, for $\varphi < 0$, not only $\tilde{\chi}(\mathbf{Q}_X)$ is enhanced, but also $\tilde{\chi}(\mathbf{Q}_Y)$ is suppressed. This suggests that nematic order promotes a transfer of magnetic spectral weight from one (inelastic) magnetic peak to the other. Such a scenario is consistent with the idea that nematic order is a manifestation of a spontaneous symmetry breaking that lifts the frustration enforced by the degeneracy of the magnetic ground state. In particular, the tetragonal symmetry at high temperatures enforce fluctuations to be equal around both ordering vectors. Once this symmetry is lowered to orthorhombic below T_S, the system selects one of the two ground states and fluctuations are enhanced only in that channel. Interestingly, inelastic neutron scattering experiments in detwinned crystals of Ni-doped $BaFe_2As_2$ observed this predicted shift of magnetic spectrum weight from one of the magnetic ordering vectors to the other [83].

In the model discussed above, it was assumed that $\tilde{\chi}_q^{-1} \propto \xi^{-2} + q^2 - i\Gamma_L\omega$, where ξ is the magnetic correlation length and Γ_L is the Landau damping. As a result, the cross sections of the spin-spin correlation function (proportional to $\text{Im}\,\tilde{\chi}_q$) around $\mathbf{Q}_X$ and $\mathbf{Q}_Y$ have a circular shape in momentum space (for a fixed energy ω). However, tetragonal symmetry allows these cross sections to be elliptical, as long as the axes of the ellipsis are rotated upon going from $\mathbf{Q}_X$ to $\mathbf{Q}_Y$. Thus, in its more general form, the effective quadratic magnetic action in the nematic state becomes:

$$\begin{aligned} S_{\text{eff}} = & \int_q \left[\xi^{-2} + (1+\eta)\, q_x^2 + (1-\eta)\, q_y^2 + \varphi\right] \boldsymbol{\Delta}_X^2 \\ & + \int_q \left[\xi^{-2} + (1-\eta)\, q_x^2 + (1+\eta)\, q_y^2 - \varphi\right] \boldsymbol{\Delta}_Y^2 \end{aligned} \tag{3.204}$$

Note that the parameter η does not break the tetragonal symmetry, but the nematic order parameter φ does. Experimentally, a nearly temperature-independent ellipticity η is observed in neutron scattering experiments [107]. If $|\eta| > 1$, higher-order momentum terms in the low-energy expansion are needed, and the magnetic ordering vector becomes incommensurate.

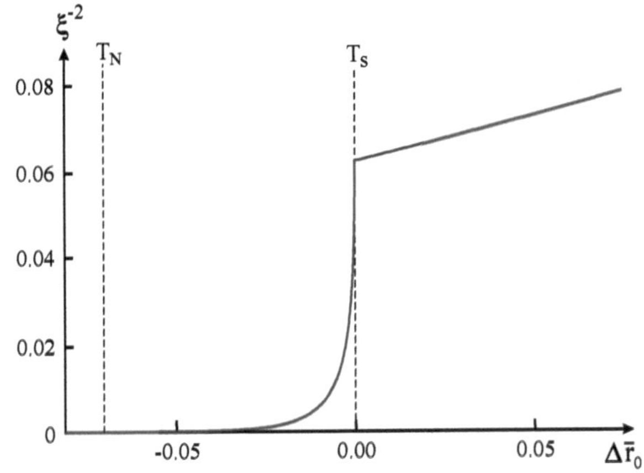

Fig. 3.13 The inverse of the squared magnetic correlation length ξ^{-2} is plotted as function of temperature using the large-N solution to the magnetic-nematic action. Here, $\Delta\bar{r}_0 \propto T - T_S$ vanishes at the nematic transition temperature. The magnetic correlation length diverges at the magnetic transition temperature T_N Adapted from [42]

The feedback of nematic order on the magnetic spectrum is not restricted to the anisotropy of the magnetic fluctuations. From the discussion above, it is clear that the renormalized magnetic correlation length (associated with the selected ordering vector) also increases according to:

$$\tilde{\xi}^{-2} = \xi^{-2} - |\varphi| \tag{3.205}$$

This enhancement of the magnetic correlation length, manifested as a discontinuity in the derivative $d\xi/dT$, is another manifestation of the degeneracy lifting promoted by long-range nematic order. In Fig. 3.13, we plot the temperature dependence of ξ obtained from the solution of the large-N equations, which displays a sharp enhancement at T_S. Such an enhancement is responsible to increase the magnetic transition temperature with respect to its hypothetical value in the tetragonal phase. Indeed, since $\xi \propto (T - T_N)^{-\nu}$, we obtain in the mean-field case ($\nu = 1/2$) an increase in the Neel transition temperature linearly proportional to the nematic order parameter:

$$\Delta T_N \propto |\varphi| \tag{3.206}$$

Experimentally, the enhancement of T_N with applied strain was observed in neutron diffraction experiments in detwinned samples [36]. The increase in the magnetic correlation length at T_S has been reported by both NMR and neutron scattering experiments in twin samples [87, 116].

3.6 Nematic Fluctuations and Pairing

We finish these notes with a few remarks on the relationship between nematicity and superconductivity. Collective nematic degrees of freedom, as discussed here, are fluctuations that behave from a symmetry point of view like B_{1g}-phonons. Note that these fluctuations are peaked at $\mathbf{q} = 0$, and therefore do not give rise to a momentum-independent pairing interaction like in the standard electron-phonon pairing problem. Being peaked at zero momentum, the pairing interaction promoted by the exchange of nematic fluctuations can favor different pairing structures in similar ways. For instance, in a simple two-band model, nematic fluctuations promote intra-band attraction and do not select between plain s-wave and extended s-wave pairing states [44, 45]. For a single-band model, it was shown in [77] that T_c is enhanced by the nematic fluctuations regardless of the symmetry of the primary pairing instability. This effect was further shown to be stronger in two-dimensional systems than in isotropic, three dimensional ones. Similar conclusions were obtained in a recent numerical analysis of a system with spin-fluctuation induced superconductivity and additional nematic fluctuations [81].

In the spin-induced nematic scenario, discussed here, nematic fluctuations are composite magnetic fluctuations and one usually expects the primary, magnetic fluctuations to couple more strongly to charge carriers. For the striped magnetic order and for systems with hole and electron Fermi surfaces, the s^{+-}-state, with opposite sign of the gap on hole and electron sheets of the Fermi surface, is a natural pairing state, even though other options with nodes of the gap exist as well [59, 101]. This does however not imply that nematic fluctuations play no role for superconductivity. First, the observation of the Raman resonance in the superconducting state demonstrates clearly that there is a sizable interaction between nematic fluctuations and the Bogoliubov quasiparticles of the superconducting state [53]. Furthermore, measurements of the elasto-resistance of several compounds revealed that optimally-doped iron-based materials generally display sizable nematic fluctuations [75]. To what extent these strong nematic fluctuations affect the pairing state of the iron-based superconductors remains an exciting topic of investigation.

Acknowledgements U.K. would like to thank the Helmholtz Association for support, through Helmholtz post-doctoral grant PD-075 "Unconventional order and superconductivity in pnictides". J.S. acknowledges the support from Deutsche Forschungsgemeinschaft (DFG) through the Priority Program SPP 1458 "Hochtemperatur-Supraleitung in Eisenpniktiden" (project-no. SCHM 1031/5-1).

We are grateful to the following colleagues for fruitful collaborations and discussions: E. Abrahams, J. Analytis, A. Baum, S. Bhattacharya, G. Blumberg, T. Böhm, A.E. Böhmer, P. C. Canfield, S. Caprara, P. Chandra, L. Chauviere, J. Chu, A.V. Chubukov, C. Di Castro, I. Eremin, I.R. Fisher, Y. Gallais, A. I. Goldman, M. Grilli, R. Hackl, F. Hardy, D. Jost, A. Kaminski, V. Keppens, M. Khodas, S. A. Kivelson, J. Knolle, H. Kontani, F. Krüger, F. Kretzschmar, S. Lederer, A. Levchenko, D. Mandrus, I. I. Mazin, M.A. McGuire, R. J. McQueeney, C. Meingast, B. Muschler, I. Paul, R. Prozorov, B.C. Sales, M. Schütt, A.S. Sefat and L.H. VanBebber. We thank R. Hackl for critical reading of large portions of the manuscript.

References

1. A. Abanov, A.V. Chubukov, Phys. Rev. Lett. **83**, 1652 (1999)
2. A. Aharony, Phys. Rev. B **8**, 3363 (1973)

3. A. Aharony, M. Fisher, Phys. Rev. B **8**, 3323 (1973)
4. A. Altland, B. Simon, *Condensed Matter Field Theory*, 2nd edn. (Cambridge University Press, Cambridge, 2010)
5. R. Applegate, R.R.P. Singh, C.C. Chen, T.P. Devereaux, Phys. Rev. B **85**, 054411 (2012)
6. S.-H. Baek, D.V. Efremov, J.M. Ok, J.S. Kim, J. van den Brink, B. Büchner, Nat. Mat. **14**, 210 (2015)
7. E.C. Blomberg, A. Kreyssig, M.A. Tanatar, R.M. Fernandes, M.G. Kim, A. Thaler, J. Schmalian, S.L. Bud'ko, P.C. Canfield, A.I. Goldman, R. Prozorov, Phys. Rev. B **85**, 144509 (2012)
8. E.C. Blomberg, M.A. Tanatar, R.M. Fernandes, B. Shen, H.-H. Wen, J. Schmalian, R. Prozorov, Nature Comm. **4**, 1914 (2013)
9. T. Böhm, A.F. Kemper, B. Moritz, F. Kretzschmar, B. Muschler, H.-M. Eiter, R. Hackl, T.P. Devereaux, D.J. Scalapino, H.-H. Wen, Phys. Rev. X **4**, 041046 (2014)
10. A.E. Böhmer, C. Meingast (2015). arXiv:1505.05120
11. A.E. Böhmer, P. Burger, F. Hardy, T. Wolf, P. Schweiss, R. Fromknecht, M. Reinecker, W. Schranz, C. Meingast, Phys. Rev. Lett. **112**, 047001 (2014)
12. A.E. Böhmer, T. Arai, F. Hardy, T. Hattori, T. Iye, T. Wolf, H.V. Löhneysen, K. Ishida, C. Meingast, Phys. Rev. Lett. **114**, 027001 (2015)
13. A.E. Böhmer, C. Meingast, C. R. Phys. **17**, 90 (2016)
14. P. Bourges, B. Keimer, S. Pailhès, L.P. Regnault, Y. Sidis, C. Ulrich, Phys. C **424**, 45 (2005)
15. M. Breitkreiz, P.M.R. Brydon, C. Timm, Phys. Rev. B **90**, 121104(R) (2014)
16. A. Cano, M. Civelli, I. Eremin, I. Paul, Phys. Rev. B **82**, 020408(R) (2010)
17. S. Caprara, C. Di Castro, M. Grilli, D. Suppa, Phys. Rev. Lett. **95**, 117004 (2005)
18. S. Caprara, C. DiCastro, T. Enss, M. Grilli, J. Magn. Magn. Mater. **321**, 686–689 (2009)
19. S. Caprara, C. Di Castro, B. Muschler, W. Prestel, R. Hackl, M. Lambacher, A. Erb, S. Komiya, Y. Ando, M. Grilli, Phys. Rev. B **84**, 054508 (2011)
20. P.M. Chaikin, T.C. Lubensky, *Principles of Condensed Matter Physics* (Cambridge University Press, Cambridge, 9 October 2000)
21. A.D. Christianson, E.A. Goremychkin, R. Osborn, S. Rosenkranz, M.D. Lumdsen, C.D. Malliakas, I.S. Todorov, H. Claus, D.Y. Chung, M.G. Kanatzidis, R.I. Bewley, T. Guidi, Nature **456**, 930 (2008)
22. J.H. Chu, J.G. Analytis, K. De Greve, P.L. McMahon, Z. Islam, Y. Yamamoto, I.R. Fisher, Science **329**, 824 (2010)
23. J.H. Chu, H.H. Kuo, J.G. Analytis, I.R. Fisher, Science **337**, 719 (2012)
24. A.V. Chubukov, S. Sachdev, J. Ye, Phys. Rev. B **49**, 11919 (1994)
25. A.V. Chubukov, D.V. Efremov, I. Eremin, Phys. Rev. B **78**, 134512 (2008)
26. A.V. Chubukov, Phys. C **469**, 640 (2009)
27. A.V. Chubukov, R.M. Fernandes, J. Schmalian, Phys. Rev. B **91**, 201105 (2015)
28. K.B. Cooper, M.P. Lilly, J.P. Eisenstein, L.N. Pfeier, K.W. West, Phys. Rev. B **65**, 241313(R) (2002)
29. R.A. Cowley, Phys. Rev. B **13**, 4887 (1976)
30. P. DeGennes, J. Prost, *The Physics of Liquid Crystals* (Clarendon, Oxford, 1993)
31. L. Dell'Anna, W. Metzner, Phys. Rev. B **73**, 045127 (2006)
32. L. Dell'Anna, W. Metzner, Phys. Rev. Lett. **98**, 136402 (2007)
33. T.P. Devereaux, D. Einzel, B. Stadlober, R. Hackl, D.H. Leach, J.J. Neumeier, Phys. Rev. Lett. **72**, 396 (1994)
34. T.P. Devereaux, R. Hackl, Rev. Mod. Phys. **79**, 175 (2007)
35. T.P. Devereaux, A.P. Kampf, Phys. Rev. B **59**, 6411 (1999)
36. C. Dhital, Z. Yamani, W. Tian, J. Zeretsky, A.S. Sefat, Z. Wang, R.J. Birgeneau, S.D. Wilson, Phys. Rev. Lett. **108**, 087001 (2012)
37. A. Dusza, A. Lucarelli, F. Pfuner, J.H. Chu, I.R. Fisher, L. Degiorgi, Europhys. Lett. **93**, 37002 (2011)
38. M. Eschrig, M.R. Norman, Phys. Rev. Lett. **85**, 3261 (2000)
39. C. Fang, H. Yao, W.F. Tsai, J.P. Hu, S.A. Kivelson, Phys. Rev. B **77**, 224509 (2008)

40. R.M. Fernandes, L.H. VanBebber, S. Bhattacharya, P. Chandra, V. Keppens, D. Mandrus, M.A. McGuire, B.C. Sales, A.S. Sefat, J. Schmalian, Phys. Rev. Lett. **105**, 157003 (2010)
41. R.M. Fernandes, E. Abrahams, J. Schmalian, Phys. Rev. Lett. **107**, 217002 (2011)
42. R.M. Fernandes, A.V. Chubukov, J. Knolle, I. Eremin, J. Schmalian, Phys. Rev. B **85**, 024534 (2012)
43. R.M. Fernandes, A.E. Böhmer, C. Meingast, J. Schmalian, Phys. Rev. Lett. **111**, 137001 (2013)
44. R.M. Fernandes, A.V. Chubukov, J. Schmalian, Nat. Phys. **10**, 97–104 (2014)
45. R.M. Fernandes, J. Schmalian, Supercond. Sci. Technol. **25**, 084005 (2012)
46. R. Folk, H. Iro, F. Schwabl, Z. Phys. B **25**(69), 371 (1976)
47. R. Folk, H. Iro, F. Schwabl, Phys. Rev. B **20**, 1229 (1979)
48. E. Fradkin, S.A. Kivelson, E. Manousakis, K. Nho, Phys. Rev. Lett. **84**, 1982–1985 (2000)
49. E. Fradkin, S.A. Kivelson, V. Oganesyan, Science **315**, 196197 (2007)
50. E. Fradkin, S.A. Kivelson, M.J. Lawler, J.P. Eisenstein, A.P. MacKenzie, Annu. Rev. Condens. Matter Phys. **1**, 153178 (2010)
51. Y. Gallais, I. Paul (2015). arXiv:1508.01319
52. Y. Gallais, R.M. Fernandes, I. Paul, L. Chauviere, Y.X. Yang, M.A. Measson, M. Cazayous, A. Sacuto, D. Colson, A. Forget, Phys. Rev. Lett. **111**, 267001 (2013)
53. Y. Gallais, I. Paul, L. Chauviere, J. Schmalian, Phys. Rev. Lett. **116**, 017001 (2016)
54. J.K. Glasbrenner, I.I. Mazin, H.O. Jeschke, P.J. Hirschfeld, R.M. Fernandes, R. Valenti, Nat. Phys. **11**, 953 (2015)
55. C.J. Halboth, W. Metzner, Phys. Rev. Lett. **85**, 5162–5165 (2000)
56. J.A. Hertz, Phys. Rev. B **14**, 1165 (1976)
57. V. Hinkov, S. Pailhès, P. Bourges, Y. Sidis, A. Ivanov, A. Kulakov, C.T. Lin, D.P. Chen, C. Bernhard, B. Keimer, Nature **430**, 650 (2004)
58. A. Hinojosa, J. Cai, A.V. Chubukov, Phys. Rev. B **93**, 075106 (2016)
59. P.J. Hirschfeld, C. R. Phys. **17**, 197 (2016)
60. R. Hlubina, T.M. Rice, Phys. Rev. B **51**, 9253 (1995)
61. D. Inosov, J.T. Park, P. Bourges, D.L. Sun, Y. Sidis, A. Schneidewind, K. Hradil, D. Haug, C.T. Lin, B. Keimer, V. Hinkov, Nat. Phys. **6**, 178 (2010)
62. T. Iye, M. Julien, H. Mayaffre, M. Horvatic, C. Berthier, K. Ishida, H. Ikeda, S. Kasahara, T. Shibauchi, Y. Matsuda, J. Phys. Soc. Jpn. **84**, 043705 (2015)
63. S. Jiang, H.S. Jeevan, J. Dong, P. Gegenwart, Phys. Rev. Lett. **110**, 067001 (2013)
64. U. Karahasanovic, J. Schmalian (2015). arXiv:1512.04861
65. U. Karahasanovic, F. Kretzschmar, T. Böhm, R. Hackl, I. Paul, Y. Gallais, J. Schmalian, Phys. Rev. B **92**, 075134 (2015)
66. S. Kasahara, H.J. Shi, K. Hashimoto, S. Tonegawa, Y. Mizukami, T. Shibauchi, K. Sugimoto, T. Fukuda, T. Terashima, A.H. Nevidomskyy, Y. Matsuda, Nature **486**, 382 (2012)
67. H.-Y. Kee, Y.B. Kim, Phys. Rev. B **71**, 184402 (2005)
68. M. Khodas, A. Levchenko, Phys. Rev. B **91**, 235119 (2015)
69. M.G. Kim, R.M. Fernandes, A. Kreyssig, J.W. Kim, A. Thaler, S.L. Bud'ko, P.C. Canfield, R.J. McQueeney, J. Schmalian, A.I. Goldman, Phys. Rev. B **83**, 134522 (2011)
70. M.V. Klein, S.B. Dierker, Phys. Rev. B **29**, 4976 (1984)
71. H. Kontani, T. Saito, S. Onari, Phys. Rev. B **84**, 024528 (2011)
72. H. Kontani, Y. Yamakawa, Phys. Rev. Lett. **113**, 047001 (2014)
73. F. Kretzschmar, T. Böhm, U. Karahasanovic, B. Muschler, A. Baum, D. Jost, J. Schmalian, S. Caprara, M. Grilli, C. Di Castro, J. Analytis, J. Chu, I.R. Fisher, R. Hackl, Nat. Phys. (2016) (in press)
74. F. Krüger, S. Kumar, J. Zaanen, J. van den Brink, Phys. Rev. B **79**, 054504 (2009)
75. H.-H. Kuo, J.-H. Chu, S.A. Kivelson, I.R. Fisher (2015). arXiv:1503.00402
76. H.-H. Kuo, J.-H. Chu, S.C. Riggs, L. Yu, P.L. McMahon, K. De Greve, Y. Yamamoto, J.G. Analytis, I.R. Fisher, Phys. Rev. B **84**, 054540 (2011)
77. S. Lederer, Y. Schattner, E. Berg, S.A. Kivelson, Phys. Rev. Lett. **114**, 097001 (2015)
78. C.C. Lee, W.G. Yin, W. Ku, Phys. Rev. Lett. **103**, 267001 (2009)

79. A.J. Leggett, Phys. Rev. **140**, A1869 (1965)
80. A.P. Levanyuk, A.A. Sobyanin, JETP **11**, 371 (1970)
81. Z.-X. Li, F. Wang, H. Yao, D.-H. Lee. arXiv:1512.04541
82. S. Liang, A. Moreo, E. Dagotto, Phys. Rev. Lett. **111**, 047004 (2013)
83. X. Lu, J.T. Park, R. Zhang, H. Luo, A.H. Nevidomskyy, Q. Si, P. Dai, Science **345**, 657 (2014)
84. M.D. Lumsden, A.D. Christianson, D. Parshall, M.B. Stone, S.E. Nagler, G.J. McDougall, H.A. Mook, K. Lokshin, T. Egami, D.L. Abernathy, E.A. Goremychkin, R. Osborn, M.A. McGuire, A.S. Sefat, R. Jin, B.C. Sales, D. Mandrus, Phys. Rev. Lett. **102**, 107005 (2009)
85. W. Lv, F. Krüger, P. Phillips, Phys. Rev. B **82**, 045125 (2010)
86. W. Lv, P. Phillips, Phys. Rev. B **84**, 174512 (2011)
87. L. Ma, G.F. Chen, D.-X. Yao, J. Zhang, S. Zhang, T.-L. Xia, W. Yu, Phys. Rev. B **83**, 132501 (2011)
88. S. Maiti, A.V. Chubukov, Phys. Rev. B **82**, 214515 (2010)
89. A.J. Millis, Phys. Rev. B **48**, 7183 (1993)
90. M. Nakajima, T. Liang, S. Ishida, Y. Tomioka, K. Kihou, C.H. Lee, A. Iyo, H. Eisaki, T. Kakeshita, T. Ito, S. Uchida, Proc. Natl. Acad. Sci. U.S.A. **108**, 12238 (2011)
91. J.W. Negele, H. Orland, *Quantum Many-Particle Systems* (Addison-Wesley, Reading, 1988)
92. V. Oganesyan, S.A. Kivelson, E. Fradkin, Phys. Rev. B **64**, 195109 (2001)
93. I. Paul, Phys. Rev. B **90**, 115102 (2014)
94. Y. Qi, C. Xu, Phys. Rev. B **80**, 094402 (2009)
95. A. Rosch, Phys. Rev. Lett. **82**, 4280 (1999)
96. E.P. Rosenthal, E.F. Andrade, C.J. Arguello, R.M. Fernandes, L.Y. Xing, X.C. Wang, C.Q. Jin, A.J. Millis, A.N. Pasupathy, Nat. Phys. **10**, 225232 (2014)
97. J. Rossat-Mignot, L. Regnault, C. Vettier, P. Bourges, P. Burlet, J. Bossy, J. Henry, G. Lapertot, Phys. C **86**, 185–189 (1991)
98. C.R. Rotundu, R.J. Birgeneau, Phys. Rev. B **84**, 092501 (2011)
99. S. Sachdev, Z. Phys. B **94**, 469 (1194)
100. S. Simayi, K. Sakano, H. Takezawa, M. Nakamura, Y. Nakanishi, K. Kihou, M. Nakajima, C. Lee, A. Iyo, H. Eisaki, S. Uchida, M. Yoshizawa, J. Phys. Soc. Jpn. **82**, 114604 (2013)
101. R. Sknepnek, G. Samolyuk, Y.-B. Lee, J. Schmalian, Phys. Rev. B **79**, 054511 (2009)
102. M.A. Tanatar, E.C. Blomberg, A. Kreyssig, M.G. Kim, N. Ni, A. Thaler, S.L. Bud'ko, P.C. Canfield, A.I. Goldman, I.I. Mazin, R. Prozorov, Phys. Rev. B **81**, 184508 (2010)
103. L. Tassini, F. Venturini, Q.-M. Zhang, R. Hackl, N. Kikugawa, T. Fujita, Phys. Rev. Lett. **95**, 117002 (2005)
104. T. Terashima et al., Phys. Rev. B **90**, 144517 (2014)
105. T. Terashima, N. Kikugawa, S. Kasahara, T. Watashige, T. Shibauchi, Y. Matsuda, A.E. Böhmer, F. Hardy, C. Meingast, H.V. Löhneysen, S. Uji, J. Phys. Soc. Jpn. **84**, 063701 (2015)
106. V.K. Thorsmolle, M. Khodas, Z.P. Yin, C. Zhang, S.V. Carr, P. Dai, G. Blumberg (2015). arXiv:1410.6456
107. G.S. Tucker, R.M. Fernandes, H.-F. Li, V. Thampy, N. Ni, D.L. Abernathy, S.L. Bud'ko, P.C. Canfield, D. Vaknin, J. Schmalian, R.J. McQueeney, Phys. Rev. B **86**, 024505 (2012)
108. F. Venturini, U. Michelucci, T.P. Devereaux, A.P. Kampf, Phys. Rev. B **62**, 15204 (2000)
109. M. Vojta, Adv. Phys. **58**, 699820 (2009)
110. M.D. Watson, T.K. Kim, A.A. Haghighirad, N.R. Davies, A. McCollam, A. Narayanan, S.F. Blake, Y.L. Chen, S. Ghannadzadeh, A.J. Schofield, M. Hoesch, C. Meingast, T. Wolf, A.I. Coldea (2015). arXiv:1502.02917
111. C. Wu, K. Sun, E. Fradkin, S.-C. Zhang, Phys. Rev. B **75**, 115103 (2007)
112. C. Xu, M. Müller, S. Sachdev, Phys. Rev. B **78**, 020501(R) (2008)
113. M. Yoshizawa et al., J. Phys. Soc. Jpn. **81**, 024604 (2012)
114. M. Zacharias, I. Paul, M. Garst, Phys. Rev. Lett. **115**, 025703 (2015)
115. W.L. Zhang, P. Richard, H. Ding, A.S. Sefat, J. Gillett, S.E. Sebastian, M. Khodas, G. Blumberg (2015). arXiv:1410.6452
116. Q. Zhang, R.M. Fernandes, J. Lamsal, J. Yan, S. Chi, G.S. Tucker, D.K. Pratt, J.W. Lynn, R.W. McCallum, P.C. Canfield, T.A. Lograsso, A.I. Goldman, D. Vaknin, R.J. McQueeney, Phys. Rev. Lett. **114**, 057001 (2015)

Chapter 4
Modeling Many-Body Physics with Slave-Spin Mean-Field: Mott and Hund's Physics in Fe-Superconductors

Luca de' Medici and Massimo Capone

Abstract Slave-spin mean-field theory, a light and accurate technique to model electronic correlations in Fermi-liquid phases of multi-orbital materials, is pedagogically exposed in this chapter, with a focus on its recent successful application to the physics of Iron-based superconductors. Beside introducing electronic correlations and recalling the Fermi-liquid phenomenology, the manuscript accompanies the step-by-step explanation of the slave-spin technique with a set of useful complements providing analytical insight into Mott and Hund's physics, which are at the heart of the physics of strongly correlated materials. Some original research material is also exposed, such as the Hund-induced shrinking of the Hubbard bands flanking the gap of the half-filled Mott insulator, and the low-energy description of the "orbital-decoupling" mechanism.

4.1 The Theoretical Description of Iron-Based Superconductors

The discovery of superconductivity with critical temperatures exceeding 50 K in the wide family of iron-based superconductors [1] has strongly impacted on the quest for the understanding of the physical origin of high-temperature superconductivity. Until this discovery, high-temperature superconductor has been for long time a synonymous of "cuprate", a chemical label which stands for the family of doped copper oxides which have been discovered by Bednorz and Müller in 1986 [2].

Lectures prepared for the XIX Training Course in the Physics of Strongly Correlated Systems, Vietri sul mare, 12–16 October, 2015.

L. de' Medici (✉)
ESPCI, Paris, France
e-mail: luca.demedici@espci.fr

M. Capone
SISSA, Trieste, Italy
e-mail: capone@sissa.it

F. Mancini and R. Citro (eds.), *The Iron Pnictide Superconductors*,
Springer Series in Solid-State Sciences 186, DOI 10.1007/978-3-319-56117-2_4

The critical temperature is not the only aspect in which the iron-based family seems to trail the copper-based relatives, as their whole phase diagram can be seen as a sort of "light" version of the celebrated doping-temperature diagram of the cuprates. In both cases a superconducting phase is established by doping a stoichiometric (parent compound) with magnetic ordering, and the two families share the existence of layers composed by the transition metal atom (Fe and Cu, respectively) even if the ligand atom in the iron-based material is not oxygen but rather a pnictogen or chalcogen atom, hence the popular label of pnictides and chalcogenides.

The first crucial difference lies in the nature of the magnetic phase from which superconductivity originates. The undoped cuprates are indeed antiferromagnetic Mott insulators, i.e., they are insulators because the strong Coulomb interaction localizes the carriers in a partially filled band, while the iron-based parents remain metallic despite the magnetic ordering in a spin-density wave pattern. This significant difference puts in jeopardy the possibility to develop a unified theoretical framework for the two classes of materials.

As a matter of fact the role of strong correlations is one of the few (almost) undisputed features of the cuprates, and it is not limited to make the parent compounds Mott insulators, but it is widely believed to be the main driving force behind the phase diagram obtained as a function of doping, including the superconducting state itself.

This suggestion, put forward by P.W. Anderson in the early days of the cuprate era [3], has deeply influenced the research in the field and drove a huge effort to understand the basic models for strongly correlated electrons, whose paradigm is the single-band Hubbard model. The combined use of a variety of techniques has confirmed that a two-dimensional Hubbard model provides at least the backbone for the theoretical description of the cuprates as doped Mott insulators.

The above mentioned chemical and structural similarities between cuprates and iron-based materials coupled with the analogy of their physics obviously led at least a part of the community to pursue the analogy also in more microscopic terms, therefore proposing that electronic correlations had to play a central role also in the iron-based material. Yet, this point of view is immediately challenged by the lack of Mott insulators in the iron-based family, which casts doubts on the strength of the electronic correlations and on their possible role in driving or affecting the superconducting phase.

This, and many other seemingly conflicting evidences led to the polarization of the community in two main camps, which we may label as weak-coupling and strong-coupling for simplicity. The first approach is corroborated by the success of Density Functional Theory (DFT) to reproduce many aspects of the bandstructure and the Fermi surface, which is typically a fingerprint of minor correlation effects. Within this approach magnetism is due to Fermi-surface nesting of otherwise metallic mobile carriers and it has an itinerant character, while superconductivity arises due to the exchange of bosons of magnetic nature [4]. The opposite approach assumes instead that the materials are close to Mott localization and they describe the magnetism as the ordering of almost localized magnetic moments. According to this picture the carriers are poorly mobile and the main interactions can be modeled in terms of a frustrated Heisenberg model with nearest- and next-nearest-neighbor superexchange interactions [5]. In the final part of this chapter we will show that a sort of intermediate scenario where both localized and delocalized carriers are simultaneously present can overcome this dualism and account for most of the properties of the iron-based phase diagram.

Before discussing the general theoretical framework of our contribution and its application to the iron-based superconductors, we briefly introduce two main physical points which motivate the outline of this chapter.

Multi-orbital Electronic Structure. One clear and significant difference between the two main families of high-temperature superconductors is the nature of the relevant electronic states. While in the cuprates only one band, mainly derived from copper $d_{x^2-y^2}$ and oxygen orbitals, crosses the Fermi surface and contributes to the low-energy physics, including superconductivity and magnetism, in the iron-based materials all the iron d orbital contribute to the low-energy states, leading to multiple Fermi surface sheets and to an inherently multiorbital fermiology.

Bandstructure Renormalization and Fermi-liquid behavior. The other aspect we would like to focus on is that experiments clearly measure well defined energy bands, even if renormalized with respect to DFT. At the same time, however, these metallic states are relatively incoherent and fragile with respect to thermal excitation, in contrast with weakly interacting metals. Therefore electronic correlations do not, for a reason or another, completely destroy the itinerant nature of the electrons even if they appear to play some role in reducing the coherence of carriers. Yet, the system remains a Fermi-liquid, even when the renormalization is quite strong.

These two basic observations lead us to focus on a class of theoretical approaches which allow to study correlated metals at a reasonable computational effort. The use of approximate methods which allow for a relatively fast exploration of phase diagrams is a crucial request for multi-orbital systems such as those required to describe iron-based superconductors. Indeed the large Hilbert space implied by the inclusion of five d orbitals severely increases the computational cost and the complexity of all the numerical or seminumerical methods one can use to solve correlated systems. This class of cheap theoretical approaches is given by slave-particle approaches, that we describe in Sect. 4.4. In particular our method of choice is the slave-spin method, which scales in a favourable way with the number of orbitals and it is therefore suitable for our goal. Importantly, this method is able to capture the quasiparticle renormalization leading to strongly correlated Fermi liquids that we aim to describe.

Of course our simplified approach cannot access all the extremely rich physics of strongly correlated materials. The main limitation is that it can only describe Fermi-liquid metallic phases, and mainly their renormalized quasiparticle structures. Only very roughly it describes the development of high-energy excitations such as the Hubbard bands or their precursors in the metallic state. In particular we have no way to properly describe a Mott insulator within this mean-field approach. On the other hand, we are able to follow the metallic solution all the way to the Mott transition.

We also limit to the paramagnetic solution without any symmetry breaking, explicitly inhibiting any magnetic ordering or superconductivity. However the modeling of the evolution of the correlation effects as a function of the interactions in the normal phase is a key information to understand how the many possible instabilities may arise at low temperature.

The chapter is organized as follows. In Sect. 4.2 we introduce the basic concepts of strong electron correlation. First we give a brief reminder of the band theory of solids followed by a basic introductions to the concepts of electronic correlation, Mott insulators and the Mott transition, and Fermi liquids. Section 4.3 presents the Hubbard model and explore its range of application and the very definition of a strongly correlated material. In Sect. 4.4 we introduce the basic concepts behind the slave-particle approaches to strongly correlated systems, and in Sect. 4.5 the slave-spin representation and its mean field approximation, with the main results on the single-band Hubbard model. In Sect. 4.6 we generalize this approach to multi-orbital systems and discuss the new energy scale intervening in these systems: Hund's coupling. Section 4.7 is dedicated to the application of all the concepts and techniques introduced in the previous section to the physics of the Iron-based superconductors and the main findings are illustrated. Finally in Sect. 4.8 the authors' concluding remarks and perspectives are given.

4.2 Strong Electronic Correlations

4.2.1 The Band Theory of Solids: A Brief Reminder

Crystalline solids are made by atoms that bind and form regular arrays or lattices. The chemical bonding and the equilibrium crystal structure of the lattice arise from the competition between the kinetic energy gained by the outermost electrons of each atom in delocalizing through the crystal, contrasted by the mutual repulsion of the ions left behind. The same delocalized outermost electrons (the "valence electrons") are responsible for most low-energy properties of the solids as they are easily excited by external fields or by thermal excitation.

Our present understanding and modeling of the electronic properties of crystalline solids, such as electrical conductivity, magnetic and optical responses for example, is founded on the "band theory" of solids [6]. Within this approach the many-body state of the huge number of electrons in the solid is constructed by populating single-particle energy levels forming a set of bands separated by energy gaps. This huge simplification is the consequence of a crucial approximation where the electron-electron Coulomb repulsion is replaced by an average effective potential experienced by each electron, thereby turning a problem of interacting electrons into a sum of independent single-electron problems. Let us briefly discuss how this can be realized in practice.

4.2.1.1 An Effective Independent-Particle Theory

In the Born–Oppenheimer approximation, in which the electronic and ionic problems are decoupled, and the ionic positions are treated as fixed parameters for the electronic problem - the many-body hamiltonian for the electrons reads:

$$\hat{H} = -\sum_i \frac{\hbar^2\nabla_i^2}{2m_e} - \sum_{i\alpha} \frac{Z_\alpha e^2}{|\mathbf{R}_\alpha - \mathbf{r}_i|} + \frac{1}{2}\sum_{i\neq i'} \frac{e^2}{|\mathbf{r}_i - \mathbf{r}_{i'}|}, \tag{4.1}$$

where $\mathbf{r}_i$ is the coordinate of electron i, $\mathbf{R}_\alpha$ the coordinate of ion α, Z_α the atomic numbers and m_e and e the mass and the charge of the electron, respectively.

Equation (4.1) defines an interacting problem which is essentially impossible to solve exactly for a realistic number of atoms. A drastic, yet surprisingly successful approximation is realized by approximating the last term with and an effective single-particle potential experienced by every single electron. With this approximation the Hamiltonian becomes separable into the sum of identical single particle hamiltonians, one for each electron, $\hat{H} = \sum_i \hat{h}(\mathbf{r}_i)$, with

$$\hat{h}(\mathbf{r}_i) \equiv \frac{\hbar^2\nabla_i^2}{2m_e} + v_{eff}(\mathbf{r}_i), \tag{4.2}$$

where $v_{eff}(\mathbf{r}_i)$ includes the two last terms in (4.1): the second, the coulomb potential due to the nuclei, treated exactly, and the third, the electronic repulsion treated in the independent electron approximation.

The next basic step is simply to solve the time-independent single-particle Schrödinger equation to obtain the single-particle eigenfunctions and eigenstates

$$\hat{h}(\mathbf{r})\phi_a(\mathbf{r}) = \epsilon_a\phi_a(\mathbf{r}). \tag{4.3}$$

For perfect crystals, where the ions sit in the positions of a Bravais lattice, as we will be considering throughout this course, $v_{eff}(\mathbf{r})$ is a function with the periodicity of the lattice, and the solutions to (4.3) will be the so-called Bloch functions of the form $\phi_a(\mathbf{r}) \equiv \phi_{\mathbf{k}n}(\mathbf{r}) = e^{i\mathbf{k}\cdot\mathbf{r}}u_{n\mathbf{k}}(\mathbf{r})$, where $u_{n\mathbf{k}}(\mathbf{r})$ is also a function with the periodicity of the lattice. These solutions are labeled by their wave number vector **k** that can take discrete values for a finite lattice with periodic boundary conditions, becoming infinitely dense for the volume of the crystal going to infinity. In practice **k** can be considered a continuous variable in macroscopic systems. On the other hand the periodicity of the lattice results in a strict equivalence between wave vectors in the first Brillouin zone of the reciprocal lattice and in the other zones (in each zone the number of allowed **k** values equals the number to the sites of the lattice), so that one usually restricts the treatment of the problem to the first zone.

In the presence of the periodic potential the eigenvalues of Bloch functions distribute in energy in distinct sets (labeled by n) within which their energy varies continuously with k following a dispersion relation $\epsilon_n(\mathbf{k}) \equiv \epsilon_{n\mathbf{k}}$. Each of these sets spans a finite energy range, a "band" of energy, separated among each others by forbidden regions of energies, the "gaps".

The many-body eigenfunction $\psi(\mathbf{x}_1, \mathbf{x}_2, \ldots)$ (where $\mathbf{x}_i \equiv \{\mathbf{r}_i, S_i^z\}$ indicates both position and spin coordinate of particle i) solving the full Hamiltonian H in the single-particle approximation will then be a product of one-body wave functions $\phi_a(\mathbf{x}_i)$.

Quantum mechanics constraints the full many body wave function for many indistinguishable fermions, like electrons are, to be antisymmetric under the exchange of any pair of particles. This means that, in the simples case of two electrons, solutions of the form $\phi_a(\mathbf{x}_1)\phi_b(\mathbf{x}_2)$ are not good physical states, and the actual state must be the antisymmetric combination

$$\psi(\mathbf{x}_1, \mathbf{x}_2) = \frac{1}{\sqrt{2}} \left[\phi_a(\mathbf{x}_1)\phi_b(\mathbf{x}_2) - \phi_b(\mathbf{x}_1)\phi_a(\mathbf{x}_2)\right]. \tag{4.4}$$

This in turn implies that $\phi_a(\mathbf{x})$ and $\phi_b(\mathbf{x})$ must be different ($a, b, \ldots$ indicate collectively all the quantum numbers characterizing a given single-particle eigenstate, thus including the spin state), or the function would vanish, and is the reason why fermions obey the Pauli exclusion principle, that states, in its most familiar formulation, that two fermions can not occupy the same quantum state.

The generalization of this construction to N particles is obtained using Slater determinants, i.e.,:

$$\psi(\mathbf{x}_1, \mathbf{x}_2, \ldots, \mathbf{x}_N) = \frac{1}{\sqrt{N!}} \begin{vmatrix} \phi_{a_1}(\mathbf{x}_1) & \phi_{a_1}(\mathbf{x}_2) & \ldots & \phi_{a_1}(\mathbf{x}_N) \\ \phi_{a_2}(\mathbf{x}_1) & \phi_{a_2}(\mathbf{x}_2) & \ldots & \phi_{a_2}(\mathbf{x}_N) \\ \vdots & \vdots & \vdots & \vdots \\ \phi_{a_N}(\mathbf{x}_1) & \phi_{a_N}(\mathbf{x}_2) & \ldots & \phi_{a_3}(\mathbf{x}_N) \end{vmatrix}, \tag{4.5}$$

which indeed ensures that the exchange of the coordinates of any two electrons results in an overall minus sign, since it implies an exchange of two columns in the determinant. Also any two identical wave functions $\phi_{a_i}(\mathbf{x}) = \phi_{a_j}(\mathbf{x})$ result in two identical lines and thus to the vanishing of the determinant, thus enforcing the Pauli principle. The energy of such wave functions is obviously $E = \sum_i \epsilon_{a_i}$.

4.2.1.2 Metals and Insulators in the Band Theory of Solids

The most celebrated success of the band theory is explaining the difference between metals and insulators [6]. Indeed the N-electron ground state of a solid is constructed using, in the Slater determinant (4.5), the N single-particle wave functions corresponding to the N lowest eigenvalues. In practice due to the spin degeneracy, we have to take only the $N/2$ lowest-energy states, each occupied twice by the two electrons with different spin directions $\{\uparrow, \downarrow\}$. The energy of highest state which is actually occupied defines the Fermi energy ϵ_F. Therefore, the electronic structure is obtained simply "filling" the levels from the bottom of the spectrum with the total number of electrons corresponding to the configuration of the individual atoms. As a result, we will end up with some completely filled bands and a (some) final one(s), called the valence band(s), which can be either partially filled or exactly filled. In the latter case, adding one more electron or exciting one of the existing electrons would have an energy cost equal to the energy gap between the valence and the next energy band, often called conduction band. In this condition the system is said to have a gap at the Fermi level.

A system with a filled valence band and empty conduction band can be shown to be inert to an applied electric field and it is therefore insulating. This can be somewhat intuitively understood by realizing that in order to create an electric current the system must be excited in a state with a finite momentum, while the equilibrium state carries no current. If the system has a gap, the excitation has a large energy cost, and the system remains in its ground state as long as the field does not reach the large values necessary to directly promote carriers from the valence to the conduction band, leading to the Landau–Zener breakdown of the insulator. On the other hand, in partially filled bands (for which the distinction between "valence" and "conduction" bands becomes meaningless, and both denominations are customarily used) finite-momentum excitations lie at a negligible distance from the Fermi energy and electrons can be excited at zero cost. Therefore band theory predicts that materials with filled bands are insulators whereas those with partially filled bands are metals.

Since each atom composing the solid brings in an integer number of electrons, in a perfect crystal the total number of electrons is a multiple of the number of lattice sites. The spin degeneracy then implies that only materials having an even number of electrons per unit cell can "fill" entirely a band. Therefore an even number of electrons per unit cell is a necessary –but not sufficient– condition to have a gap at the Fermi level, in band theory. On the other hand systems with an odd number of electrons per unit cell will always end up with partially filled bands. Indeed all elemental insulators that one finds by browsing the periodic table have an even number of electrons per unit cell. If the band gap in an insulator is smaller than $k_B T$ (where k_B is the Boltzmann constant and T the absolute temperature) at ambient temperature, thermal processes can excite electrons across the gap opening a conduction channels. These systems are called semiconductors. Another general prediction of band theory is that each time the population of a band insulator is modified either by chemical doping or by electrostatic gating, either the conduction or the valence band necessarily become partially populated and the system becomes a metal.

4.2.1.3 Fermi Velocity and Band Mass

Within the framework of the band theory, all the electrons in a solid behave as independent particles, but the parabolic dispersion of electron in the vacuum $\epsilon_k = \hbar^2 k^2 / m_e$ is replaced by a series of band dispersions $\epsilon_{n\mathbf{k}}$ determined by the periodic potential of the ionic lattice and by the effective potential arising from the electron-electron interactions.

The band dispersion is therefore largely responsible of the electronic properties of a solid. In particular the electrons near to the Fermi energy are responsible for most physical responses of a given material. It is particularly useful to introduce effective quantities relating the behavior of band electrons with electrons in free space.

By linearizing the dispersion close to the Fermi energy we can define a band Fermi velocity

$$\mathbf{v}_{n\mathbf{k}} = \frac{1}{\hbar}\nabla_{\mathbf{k}}\epsilon_{n\mathbf{k}}. \tag{4.6}$$

This quantity will be obviously different from that of a non-interacting electron in vacuum $|\mathbf{v}_{n\mathbf{k}}| = \hbar k/m_e$, and we can use it to define a band effective mass through the relation $|\mathbf{v}_{n\mathbf{k}}| = \hbar k/m_b$, i.e., the same expression of free electrons where the bare mass m_e is replaced by an effective mass m_b, which measures the inertia of the bands electrons. Notice that this definition is unambiguous only close to band minima, where the dispersion is quadratic and the Fermi velocity is actually proportional to the momentum, while in general cases different definitions can be used according to the context.

The band mass is however a very useful parameter which characterizes how some experimental quantities differ in band theory from the free-electron value: for instance the low-temperature electronic specific heat in metals is typically altered by a factor $\sim m_b/m_e$ with respect to a Fermi gas and similar behavior is found for response functions such as the spin susceptibility.

4.2.2 Electron–Electron Interactions and Correlations

Band theory is so successful in general that is regarded as a sort of "standard model" of the solid state. While many qualitative predictions can be based on simple analysis of the chemistry of materials, namely on the electron count and the consequent filling of the bands, in order to obtain a quantitative theory able to capture the differences between different materials, the choice of the effective one-electron potential $v_{eff}(\mathbf{r})$ becomes crucial but is also highly non trivial. The most popular and successful method to derive $v_{eff}(\mathbf{r})$ starting from the actual Coulomb interactions is Kohn–Sham Density-Functional Theory (DFT), where effective Kohn–Sham single-particle orbitals and the effective potential are self-consistently determined [7]. DFT requires in turn a choice of the so-called exchange-correlation functional, which defines possible "flavours", among which the local density approximation (LDA) and generalized gradient approximation (GGA) play a prominent role in actual implementations. Alternative single-particle potentials can be defined, for example through a Hartree–Fock variational principle which defines an effective single-particle description alternative to Kohn–Sham DFT. As a matter of principle, all the methods that derive describe the electronic structure of a solid in terms of a collection of well defined bands are based on the definition of some kind or explicit of implicit effective potential $v_{eff}(\mathbf{r})$.

There are however important cases in which even the most powerful band theory calculations are grossly inaccurate, or fail even at the most qualitative level, misrepresenting as good metals materials which are experimentally insulating. This happens when the effects of electron-electron interactions are so relevant that the independent-electron approximation of band theory fails to correctly account for the physics of the system. Broadly speaking, the valence electrons of these materials are said to be "strongly correlated", a label which is often extended to the materials.

For an oversimplified yet transparent picture of how strong correlations imply a failure of the independent-electron scenario we can consider a simple two-electron problem with two "lattice sites", which can be thought as ions of an H^2 molecule. We assume that the electrons can only occupy wave functions localized around one of the two sites, say $\phi_L(\mathbf{r})$ and $\phi_R(\mathbf{r})$. Within the independent-electron approximation the wave function of the electron will be a linear combination of the two basis states with normalized coefficients: $\phi_a(\mathbf{x}) = c_{aL}\phi_L(\mathbf{x}) + c_{aR}\phi_R(\mathbf{x})$. The 2-particle function will then be the antisymmetrized sum (4.4) of two products of the kind:

$$\begin{aligned}\phi_a(\mathbf{x}_1)\phi_b(\mathbf{x}_2) &= c_{aL}c_{bL}\phi_L(\mathbf{x}_1)\phi_L(\mathbf{x}_2) + c_{aR}c_{bR}\phi_R(\mathbf{x}_1)\phi_R(\mathbf{x}_2)\\ &+ c_{aL}c_{bR}\phi_L(\mathbf{x}_1)\phi_R(\mathbf{x}_2) + c_{aR}c_{bL}\phi_R(\mathbf{x}_1)\phi_L(\mathbf{x}_2))\end{aligned} \tag{4.7}$$

If the relative effect of the repulsive electron-electron interaction (which depends on the specific shape and distance of the basis functions, on possible external screening, etc.) is important, it is easy to understand that in the exact solution the first two terms in (4.7) in which the two electrons are close to one another and thus maximize the energy due to mutual repulsion must have a smaller weight than the last two. However direct inspection of the coefficient leads to realize that there is *no way* for the independent electron approximation to account for this effect, since there is no way to depress the relative contribution of the first two terms in the sum by acting on the coefficient of the single-particle solutions.

The reason for the failure of the independent-electron picture becomes now clear. If the interactions are strong the electrons pay a large energy cost if they are on the same site (or in general if they are close) or, in other words, they gain energy in *avoiding* each other. As a consequence the many-electron wave function is depressed when one or more electronic coordinates are at small reciprocal distances, i.e., when $\mathbf{r}_i - \mathbf{r}_j$ is small with respect to the effective range of interactions. Therefore the behavior of each electron depends on the state of the other and a single-particle description breaks down. This is what we define as a *correlated* behavior, which can not be encoded in a Slater determinant, or equivalently in a band-theory description.

4.2.3 The Mott Transition and Mott Insulators

If we imagine to continuously vary the degree of correlation in our system, for example changing the inter-site distance in our two-site example, we expect an evolution from a "weakly interacting" regime (small distances) where the single-particle description is a fair approximation of the exact solution and a "strongly interacting" regime, where the correlation between the electrons dominates and the independent electron picture breaks down.

Let us consider an idealized system with a single band and a number of electrons equal to the number of lattice sites. In the weakly interacting regime, where a single-particle picture holds, the solution of our system describes a metal with a half-filled band due to the odd number of electrons per idealized atom. When the interaction is increased, which can be realized increasing the lattice spacing as in our two-site example, the electrons will become more and more correlated in the sense we described above.

This process where the correlation between the electrons increases is characterized by the competition between the kinetic energy, which favors delocalized band states and the interaction, which disfavors configurations with two electrons on the same site, which determines constraints to the motion of the electrons. When the energetic cost of a doubly occupied site exceeds the possible energy gain, the best compromise to minimize the energy is simply to localize the electrons, one on each site. This avoids to pay an energy of order U, even if the kinetic energy is obviously sacrificed. Therefore the systems becomes an insulator, despite the electron count would imply a metal according to the band theory of solids. This quantum state, where the electrons are localized by the mutual Coulomb repulsion, is called a Mott insulator after N. Mott, who introduced the concept in a pioneering work [8]. The transition from a metal to a Mott insulator which occurs as a function of the ratio between Coulomb repulsion and hopping is called a Mott transition. It is important to notice that Mott localization can happen only when the number of electrons equals the number of sites, or in multi orbital systems, for an integer number of electrons per atom. If the system is doped away from half-filling, electronic correlation can have a major effect, but they can not drive a Mott insulating state in the absence of symmetry breaking. Therefore doping of a Mott insulator in principle leads to a metallization. As a matter of fact, one can drive an insulator-to-metal transition by doping a Mott insulator with either holes or electrons.

Needless to say, the very existence of Mott insulators represents a clear failure of the band theory of solids. Indeed Mott insulators have been experimentally discovered already in 1937 [9], and a number of these materials have been identified in the years including related charge-transfer insulators [10], whose prominent members are the parent compounds of high-temperature superconductors. The physics of high-temperature superconductivity is indeed widely believed to be the consequence of the doping of a (charge-transfer) Mott insulator.

It must also be mentioned that Mott insulators are usually characterized by long-range ordering of the spin (and orbital) degrees of freedom. It is indeed easy to show that a half-filled Hubbard model is unstable towards two-sublattice antiferromagnetism. In the case where the hopping is limited to nearest-neighbors sites only, the Hubbard model has indeed an antiferromagnetic ground state for every value of U, while longer-range hoppings lead to a critical value of the interaction. The long-range ordered phase can survive for a finite range of doping thereby partially hiding the intrinsic Mott transition. In this work we will focus only on paramagnetic solutions, where the long-range magnetic ordering is inhibited. This choice allows to follow the intrinsic effects of electron-electron correlations, disentangling them from the appearance of low-energy broken symmetry phases.

4.2.4 Fermi-Liquids and Effective Mass

The very existence of Mott insulators and the Mott transition are qualitative and unequivocal fingerprints of strong correlations leading to a complete breakdown of any independent-electron approximation. Yet, many materials display strong correlations but remain metallic, either because their interactions are not sufficient to drive Mott localization or simply because the electron count is not commensurate with the number of sites. Interestingly, the behavior of many correlated metals is not qualitatively different from weakly interacting metals. As a matter of fact the low temperature behavior of many observables has the same temperature dependence of free electrons only with different parameters, as described for example by Drude–Sommerfeld theory, even in very strongly correlated materials on the brink of Mott localization.

The robustness of the metallic behavior in interacting fermionic systems is described by the Landau theory of normal Fermi liquids [11–13]. The basic assumption of this phenomenological theory is that the low-energy and low-temperature elementary excitation of a system of interacting fermions are in one-to-one correspondence with the excitations of a system of non-interacting particles, as if one could adiabatically switch on the interaction without changing the character of the excitations.

More concretely, the excitations of a non-interacting Fermi gas are given by variations of the occupations of the different single-particle eigenstates and they are associated with the creation and/or destruction of actual fermions. Analogously, the excitations of the interacting systems correspond to "quasiparticle states", excitations with fermionic statistics which carry the same quantum numbers of the original fermions. Obviously, these states are not real eigenstates of the interacting systems and the crucial idea is that the above correspondence is only valid for weak excitations at low-energy. In general a quasiparticle excitation has a finite lifetime, as opposed to non-interacting particles. The Landau theory is based on the fact that the inverse of the lifetime vanishes quadratically both as a function of the distance from the fermi Energy and as a function of temperature

$$\frac{1}{\tau} = a(\epsilon - \epsilon_F)^2 + bT^2 \tag{4.8}$$

Notice that no assumption is made about the strength of the interactions, that can well be large, as long as it does not lead to a complete breakdown of the metallic state. On the other hand, the residual interactions between quasiparticles will be small. The whole Fermi liquid theory essentially amounts to phenomenologically describe a system of strongly interacting particles in terms of a low-energy picture of renormalized quasiparticles with small residual interactions. The effect of the original interactions is therefore included both in the residual interactions and in effective parameters characterizing the quasiparticle states. Among these, a central role is played by the effective mass, which emerges in a similar way as in band theory in terms of an effective renormalized dispersion $\epsilon^*_{n\mathbf{k}}$ –which now arises also from interaction effects– and a renormalized Fermi velocity $\mathbf{v}^*_{n\mathbf{k}} = \frac{1}{\hbar}\nabla_{\mathbf{k}}\epsilon^*_{n\mathbf{k}}$. The effective mass m^* is now defined by the relation

$$|\mathbf{v}^*_{n\mathbf{k}}| = \frac{\hbar k}{m^*} \tag{4.9}$$

The physical meaning of the effective mass is the same of the band effective mass, namely it is an attempt to compare the motion of the interacting electron with that of non-interacting electrons. In this context, where we focus on the interaction effects, it is natural to compare the effective mass with the band mass rather than with the free-electron mass, and the ratio m^*/m_b can be used as an estimate of the degree of correlation of a given system.

This effective mass describes the reduction of the mobility of the electrons as a consequence of the interactions that we discussed above. The whole process of Mott localization that we described in the above section can be mirrored in a progressive enhancement of the effective mass from the non-interacting value to a divergent value which corresponds to the localization of the electron. The effective mass in strongly interacting materials can range from a few times the bare mass to huge values of the order of hundred times the bare value, which correspond to the so-called heavy fermion materials [14].

As long as the system remains in a Fermi-liquid state, i.e., it does not turn into a Mott insulator, into a state with some kind of ordering (magnetic, orbital, ...) or into a more exotic non-Fermi liquid metal, the effective mass embodies the main effect of correlations and it can be used to understand the influence of interactions on a given system. It should be mentioned that also other interaction mechanism, as e.g. any kind of electron-boson interaction, also lead to an analogous enhancement of the effective mass.

The Landau Fermi-liquid theory is a phenomenological approach. However it can be founded on a diagrammatic perturbative expansion of the interaction. The condition (4.8) is verified when the imaginary part of the self-energy $\Sigma(\omega)$ vanishes like ω^2 for small frequencies and like T^2 for small temperatures. Focusing on the zero-temperature behavior, up to order ω the self-energy is purely real and it can be expanded as

$$\Sigma(\mathbf{k},\omega) \simeq \Sigma'(\mathbf{k},0) + \omega\frac{\partial\Sigma'(\mathbf{k},\omega)}{\partial\omega}|_0, \tag{4.10}$$

where $\Sigma'(\mathbf{k},\omega)$ is the real part of the self-energy. The Green's function then becomes

$$G(\mathbf{k},\omega) \simeq \frac{1}{\omega - \epsilon_{\mathbf{k}} + \mu - \Sigma'(0) - \omega \left.\frac{\partial \Sigma'(\mathbf{k},\omega)}{\partial \omega}\right|_0} = \frac{1}{\omega\left(1 - \left.\frac{\partial \Sigma'(\mathbf{k},\omega)}{\partial \omega}\right|_0\right) - \epsilon'_{\mathbf{k}} + \mu}$$
$$= \frac{1}{\omega/Z_{\mathbf{k}} - \epsilon'_{\mathbf{k}} + \mu} = \frac{Z_{\mathbf{k}}}{\omega - Z_{\mathbf{k}}(\epsilon'_{\mathbf{k}} - \mu)}, \tag{4.11}$$

where we defined a shifted dispersion $\epsilon'_{\mathbf{k}} = \epsilon_{\mathbf{k}} + \Sigma'(\mathbf{k}, 0)$ and the quasiparticle weight

$$Z_{\mathbf{k}} = \frac{1}{1 - \left.\frac{\partial \Sigma'(\mathbf{k},\omega)}{\partial \omega}\right|_0}. \tag{4.12}$$

The final result shows that the renormalized Green's function features a pole with a weight $Z_{\mathbf{k}}$ (hence the name of quasiparticle weight) at a frequency $\omega = Z_{\mathbf{k}}(\epsilon'_{\mathbf{k}} - \mu)$, which plays the role of a renormalized quasiparticle energy. If we further expand the self-energy and the dispersion around the Fermi momentum, we obtain the expression for the effective mass

$$\frac{m_e}{m*} = \frac{1 - \left.\frac{\partial \Sigma'(\mathbf{k},0)}{\partial k}\right|_{k_F}}{1 - \left.\frac{\partial \Sigma'(\mathbf{k}_F,\omega)}{\partial \omega}\right|_0}. \tag{4.13}$$

We notice that if the self-energy does not depend on momentum, as it will happen in the mean-field approaches that we will discuss in the present chapter and for the Dynamical Mean-Field theory where the spatial dependence of the observables is frozen, the numerator of (4.13) becomes 1 and we find

$$\frac{m*}{m_e} = \frac{1}{Z}, \tag{4.14}$$

so that the effective mass is just given by the inverse of the quasiparticle weight. An increased effect of correlation will be then reflected in the reduction of Z and a metal-insulator transition will occur when Z vanishes or, equivalently, the effective mass diverges.

The Landau theory of Fermi liquids focuses on the low-energy excitations of a metallic system and it does not make any assumption for the high-energy and high-temperature behavior, which could be accessible through the knowledge of the self-energy at all frequencies. The methods and the results presented in this chapter indeed focus on strongly correlated Fermi liquids which can be described within the above formalism, and will not address possible violations of the Fermi liquid as well as the development of the high-energy Hubbard bands. This choice is motivated and justified by the fact that the metallic phases of iron-based superconductor appear experimentally as regular Fermi liquids with only minor exceptions.

4.3 The Hubbard Model

In this section we introduce the celebrated Hubbard model, which is almost universally considered the paradigmatic model to study the physics of strong correlations. The model has been introduced independently by Hubbard, Kanamori and Gutzwiller to study itinerant magnetism, but later became a general reference system to study correlated metals and their metal-insulator transitions. As we will see, the model can be derived as a simplification of the many-body Hamiltonian describing electrons in a solid, which allows to treat explicitly an essential part of electron-electron correlation. In second quantization the many-body hamiltonian for the electrons (4.1) reads $\hat{H} = \hat{H}_0 + \hat{H}_{int}$

$$\hat{H}_0 = \sum_{\sigma} \int d\mathbf{r} \Psi^{\dagger}_{\sigma}(\mathbf{r}) \left[-\frac{\hbar^2 \nabla^2}{2m_e} + V(\mathbf{r}) \right] \Psi_{\sigma}(\mathbf{r}) \tag{4.15}$$

where $V(\mathbf{r}) = -\sum_{\alpha} \frac{Z_{\alpha} e^2}{|\mathbf{R}_{\alpha} - \mathbf{r}|}$, and

$$\hat{H}_{int} = \sum_{\sigma} \sum_{\sigma'} \int d\mathbf{r} d\mathbf{r}' \Psi^{\dagger}_{\sigma}(\mathbf{r}) \Psi^{\dagger}_{\sigma'}(\mathbf{r}') \left[\frac{1}{2} \frac{e^2}{|\mathbf{r} - \mathbf{r}'|} \right] \Psi_{\sigma'}(\mathbf{r}') \Psi_{\sigma}(\mathbf{r}) \tag{4.16}$$

where $\Psi_{\sigma}(\mathbf{r})$ is the field operator creating a particle with spin σ at point $\mathbf{r}$. H_0 is diagonalized by bloch wavefunctions $\phi_{\mathbf{k}n}(\mathbf{r})$. These are delocalized through the solid and to implement any sort of approximation involving the interaction term it is convenient to re-express them using a complete orthonormal basis of localized functions, like a basis of Wannier orbitals [6], for example (but not uniquely [15]) defined by:

$$w_n(\mathbf{r} - \mathbf{R}) = \frac{1}{\sqrt{N}} \sum_{\mathbf{k} \in BZ} e^{-i\mathbf{k} \cdot \mathbf{R}} \phi_{\mathbf{k}n}(\mathbf{r}), \tag{4.17}$$

where a Wannier orbital is associated to each lattice position $\mathbf{R}$. This choice is justified and particularly useful in strongly correlated materials because, as discussed in Sect. 4.3.2, strong correlations emerge when the conduction electron density is very concentrated around the ionic positions. The field operator can be then expressed in this basis:

$$\Psi^{\dagger}_{\sigma}(\mathbf{r}) = \sum_{in} w^*_n(\mathbf{r} - \mathbf{R}_i) d^{\dagger}_{in\sigma} \tag{4.18}$$

(where we used the index to label the site with position R_i). Now $d^{\dagger}_{in\sigma}$ creates an electron with spin σ in the n-th Wannier orbital associated to site R_i. In this basis the Hamiltonian reads[1]

[1]Here, and throughout this manuscript, we use light (i.e., non-bold) characters for the site indices $i, j, \ldots$. Indeed each of these can be seen formally as a scalar discrete index labeling all the sites.

$$\hat{H} = \sum_{ijmm'\sigma} t_{ij}^{mm'} d_{im\sigma}^{\dagger} d_{jm'\sigma} + \frac{1}{2} \sum_{ijkl} \sum_{mm'nn'} \sum_{\sigma\sigma'} V_{ijkl}^{mm'nn'} d_{im\sigma}^{\dagger} d_{jm'\sigma}^{\dagger} d_{kn'\sigma'} d_{ln\sigma} \quad (4.19)$$

where

$$t_{ij}^{mm'} = \int d\mathbf{r}\, w_m^*(\mathbf{r} - \mathbf{R}_i) \left[\frac{-\hbar^2 \nabla^2}{2m_e} + V(\mathbf{r}) \right] w_{m'}(\mathbf{r} - \mathbf{R}_j) \quad (4.20)$$

and

$$V_{ijkl}^{mm'nn'} = \int d\mathbf{r} d\mathbf{r}' w_m^*(\mathbf{r} - \mathbf{R}_i) w_{m'}^*(\mathbf{r}' - \mathbf{R}_j) \frac{e^2}{|\mathbf{r} - \mathbf{r}'|} w_n(\mathbf{r}' - \mathbf{R}_k) w_{n'}(\mathbf{r} - \mathbf{R}_l) \quad (4.21)$$

Up to here the only approximation performed is the Born–Oppenheimer one and we have merely rewritten the many-body problem into a Wannier basis, which emphasizes the real space coordinates of the ions. A typical treatment of the many-body Hamiltonian uses band theory as a starting point, i.e., uses the basis of Bloch functions that diagonalize the effective problem in which $V_{eff}(\mathbf{r})$ is used in (4.15) instead of $V(\mathbf{r})$ (as explained in Sect. 4.2), and the corresponding basis of Wannier orbitals. This means including then the static electrostatic potential due to the electrons in H_0 and will obviously require to subtract it during the treatment of (4.16) to avoid a "double counting" of this term.

The convenience of the Wannier basis in strongly correlated materials stems out of the fact that when the basis orbitals are very localized in space the hopping integral $t_{ij}^{mm'}$ decays rapidly with the distance between the sites i and j, and thus the number of independent parameters for the one-body part of the Hamiltonian is reduced, and the corresponding Hamiltonian matrix rather sparse.

This is true also for the interaction parameters $V_{ijkl}^{mm'nn'}$ which indeed decay faster with the distance because of the spatial decay of the Coulomb interaction and because they involve products of four wave functions instead of two.

Even with a wise choice of the set of basis functions the problem at hand remains formidably complex and simplifications are needed. Two key physical facts are to be invoked here. First, when analyzing the properties of correlated materials at the temperatures and for most of the probes relevant to experiments[2] only a subset of bands is relevant. Indeed those coming from deep shells of very low energy are completely filled and can be considered inert to weak perturbations, like empty bands far above the Fermi energy. It is thus reasonably safe to limit the explicit treatment of many-body effect to a subset of bands (i.e., limiting the orbitals m in the sum) near the Fermi energy. Second, these degrees of freedom can be eliminated modulo the inclusion of the screening effects on the interaction between the remaining electrons. This amounts to replacing in the integrals (4.21) the coulomb interaction $e^2/(|\mathbf{r} - \mathbf{r}'|)$

(Footnote 1 continued)
It is however more natural to think about them as vectors of 3 discrete indices, on a 3-dimensional lattice, indeed.

[2]With the notable exception of high-energy probes like e.g. X-rays that involve states vary far from the Fermi energy.

with the screened interaction $W(|\mathbf{r}-\mathbf{r}'|)$ that decays much faster with distance.[3] It is to be noted that only the screening from the eliminated degrees of freedom has to be included, since the explicit treatment of the interactions in the retained subset of bands will account for the rest of it.

This simplification allows then a clear hierarchy of the interaction integrals, those for which all the Wannier functions are on the same site being much larger than those involving other sites. One customarily (but indeed not always) retains only the former. i.e., the local interaction

$$V_{ijkl}^{mm'nn'} = U^{mm'nn'}\delta_{ij}\delta_{ik}\delta_{il} \tag{4.22}$$

to obtain what is called a (multi-orbital) Hubbard model, where the interactions are local, but involve electrons in different orbitals. When we consider just one orbital on every site we obtain the single-band Hubbard model, reading:

$$\hat{H} = \sum_{ij\sigma} t_{ij} d^{\dagger}_{i\sigma} d_{j\sigma} + U \sum_{i} n_{i\uparrow} n_{i\downarrow}, \tag{4.23}$$

where we introduced the number operators $n_{i\sigma} \equiv d^{\dagger}_{i\sigma} d_{i\sigma}$.

The Hubbard model is clearly a huge simplification of the original problem, and in many realistic systems we can even speak of an oversimplification.[4] Yet, it retains the basic aspects of the strong correlation physics and it includes the essential physics behind Mott localization and formation of bad metals with large effective masses. For this reason the model remains very hard to solve and it is unsolved except in one or infinite spatial dimensions or in extreme and trivial limits $t_{ij} = 0$ or $U = 0$.

4.3.1 Complement: *Single-Band Hubbard Model at Particle-Hole Symmetry*

Here we show that, for the single-band Hubbard model on a bipartite lattice (i.e., a lattice that can be divided in two sublattices A and B such that the electrons hop only from one sublattice to the other, the simplest example being a square lattice with hopping only between nearest neighbours), the choice $\mu = U/2$ fixes the filling to one electron per site. This corresponds to particle-hole symmetry in this model, and this result is thus independent of the value of U and of the temperature, and holds irrespectively of the phase in which the system is (metallic, insulating, ..).

Indeed if we apply to (4.23) the particle-hole transformation:

---

[3]The simplest analytical form is the Yukawa potential: $W(|\mathbf{r}-\mathbf{r}'|) = e^2 \frac{e^{-\lambda|\mathbf{r}-\mathbf{r}'|}}{|\mathbf{r}-\mathbf{r}'|}$, where λ is the screening length.

[4]Despite its simplicity, this is not necessarily too crude an approximation, as it is believed to describe, at least qualitatively, the main physics of the copper-oxygen planes of the cuprates, and it can be simulated with fermionic ultra cold atoms trapped in optical lattices.

$$\begin{aligned} d^{\dagger}_{i\sigma} &\rightarrow d_{i\bar{\sigma}} \\ d_{i\sigma} &\rightarrow d^{\dagger}_{i\bar{\sigma}} \end{aligned} \tag{4.24}$$

the Hamiltonian becomes

$$\hat{H} - \mu\hat{N} = -\sum_{ij\sigma} t_{ij} d^{\dagger}_{i\sigma} d_{j\sigma} + U \sum_{i} n_{i\uparrow} n_{i\downarrow} + (\mu - U) \sum_{i\sigma} n_{i\sigma} + U - 2\mu, \tag{4.25}$$

in which $t_{ij} = t_{ji}$ (hermiticity of the Hamiltonian) was used when renaming the site indices. One can then apply a gauge transformation of the fermionic fields only on one sublattice,

$$\begin{aligned} d^{\dagger}_{i\sigma} &\rightarrow -d_{i\sigma} \\ d_{i\sigma} &\rightarrow -d^{\dagger}_{i\sigma} \\ i &\in \mathit{sublattice}\ A \end{aligned} \tag{4.26}$$

to obtain

$$\hat{H} - \mu\hat{N} = \sum_{ij\sigma} t_{ij} d^{\dagger}_{i\sigma} d_{j\sigma} + U \sum_{i} n_{i\uparrow} n_{i\downarrow} + (\mu - U) \sum_{i\sigma} n_{i\sigma} + U - 2\mu, \tag{4.27}$$

a Hamiltonian identical to $\hat{H} - \mu\hat{N}$ with $\hat{H}$ as in (4.23) if the chemical potential has the value $\mu = U/2$.

Thus $\mu = U/2$ enforces the particle-hole symmetry and hence $\langle n_{i\uparrow} + n_{i\downarrow} \rangle = 1$, i.e., half-filling.

It can be convenient to rewrite the Hubbard Hamiltonian in a manifestly invariant form by rescaling the chemical potential. Indeed completely equivalent forms for the interaction to (4.23) are both

$$U \sum_{i} \left(n_{i\uparrow} - \frac{1}{2}\right)\left(n_{i\downarrow} - \frac{1}{2}\right) \tag{4.28}$$

and

$$\frac{U}{2} \sum_{i} (n_{i\uparrow} + n_{i\downarrow} - 1)^2 = \frac{U}{2} \sum_{i} \left(\sum_{\sigma} \left(n^{d}_{i\sigma} - \frac{1}{2}\right)\right)^2 \tag{4.29}$$

in which the particle-hole symmetry is evident. Using these forms of interaction the chemical potential yielding half-filling will then be $\mu = 0$. By developing the above forms[5] one can convince himself that this corresponds indeed to the standard form (4.23) with $\mu = U/2$.

---

[5]Here one uses the fact that for fermions $n^2_{i\sigma} = n_{i\sigma}$.

4.3.2 Which Materials Are Likely to Be Strongly Correlated?

In the simple example of Sect. 4.2.2, we have shown that "strong correlations" appear when some short-range Coulomb interaction exceeds the energy gain in freely delocalizing the carriers, thus making configurations with two electrons on the same atom unfavourable. In this section we give some basic guidelines to understand how the balance between these two energy terms changes in different materials, and in particular we identify which compounds are more likely to display strong electronic correlations.

Within the tight-binding parametrization leading to the Hubbard modeling described in Sect. 4.3 the delocalizing energy is given by the hopping matrix elements $t_{ij}^{mm'}$, which have to be compared with the interaction parameters $U^{mm'nn'}$. A material is expected to be more or less correlated according to the ratio between the Hubbard U and some combination of the hopping matrix elements.

As we anticipated, we focus on the valence electrons, and we expect that the kinetic energy is mainly determined by the largest hoppings, which naturally correspond to nearest-neighboring ions due to the rapid decay of the overlap between wavefunctions when the distance increases. Roughly speaking, the hopping is determined by the overlap between valence wavefunctions associated with neighbouring lattice sites. This quantity is expected to depend on the spatial extension of the relevant wavefunctions, which naturally depends on their atomic quantum numbers n, l and m_l and on the value of the lattice spacing, which also depends on the chemistry of the materials and on the atomic species involved.

On the other hand, the Coulomb interaction is a local quantity, which depends much less on the chemistry of the material and the orbitals involved. Of course this last statement has to be taken with a grain of salt, as U changes in the various materials, mainly because of the different screening, but certainly the leading effect in determining the degree of correlation in a materials is the variation of the hopping matrix elements.

With these guidelines, we can understand why certain families of materials are more likely to display strong correlation effects than others. As we already mentioned, the key factor is given by the symmetry of the orbitals which mainly contribute to the valence bands. The hopping matrix elements are controlled by the lattice spacing and the spatial extension of the wavefunctions. As a matter of fact the critical parameter to determine the degree of correlation of a materials is the latter, which depends directly on the nature of the involved orbitals, while the lattice spacing has a weaker and much less obvious dependence on the atomic number and it is influenced by the core electrons.

On the other hand, the spatial extension of the valence wavefunctions essentially increases as the atomic number grows, as we know from basic quantum mechanics. The reason for this is that each wavefunction must be orthogonal to all the previous ones, which is realized by increasing the number of nodes and by shifting the charge further away from the nucleus. Based on this considerations, it would be hard to find a general trend for the evolution of the hopping (hence of the degree of correlation) as a function of the atomic number of the atoms involved. There is however a crucial aspect to take into account. As it is well known the orbital quantum number l can assume integer values smaller than the principal quantum number n. Therefore, at each value of n the orbitals with $l = n - 1$ are automatically orthogonal to all the "previous" orbitals because they are the first with that value of l. This implies that these orbitals do not need to push the charge far from the nucleus, and they are more localized around the lattice sites. This leads to relatively small value of the hopping. Since the Hubbard U is not equally affected, this results in an increased level of correlation.

Hence, we can expect strong correlation effects are particularly strong in materials where the valence bands arise from 3d orbitals, the first orbital with $l = 2$ and the 4f orbitals, the first with $l = 3$. This expectation is actually met, and the prototype materials for strong correlation physics are characterized by the presence of transition-metal atoms or rare earths, the two groups which correspond to the above quantum numbers. As a matter of fact the 4f are more correlated than the 3d, and display extreme correlation effects (e.g. the huge effective masses of heavy-fermion compounds), while the 3d happen to be exactly in the "interesting" range, and they can either be strongly correlated metals or Mott insulators according to different chemical and structural factors. For the same reason 3d materials can be driven across the Mott metal insulator transition either by doping or by pressure and chemical substitution and they represent the ideal playground to study and understand the properties of strongly correlated materials. Besides the academic interest, these materials also display a wealth of spectacular phenomena, including high-temperature superconductivity since both iron and copper belong to the 3d transition metal series.

4.4 Slave-Particle Approaches

In the previous sections of this chapter we have emphasized that the physics of electron-electron correlations driving the Mott transition can not be described within *any* single-particle picture. This is true also for the Hartree–Fock approximation, which amounts to search for the optimal Slater determinant for a given problem according to the variational principle, which obviously corresponds to an effective single-particle picture. Since the Hartree–Fock method is completely equivalent to a mean-field decoupling of the interaction, any attempt to theoretically account for strong correlation physics requires to go beyond the standard mean field.

The need to understand the effects of correlations, primarily triggered by the discovery of high-temperature superconductivity, led to the development of methods able to treat correlated systems, ranging from purely numerical methods such as quantum Monte Carlo simulations or the Density-Matrix Renormalization Group to analytical approximate approaches like the functional renormalization group, to mention only a few members of a very long list [16–19].

In this contribution we will focus on a class of methods which are based on the introduction of auxiliary "slave" particles. These approaches allow to construct a new kind of mean-field theories specifically designed to deal with strongly correlated systems and to describe the Mott–Hubbard transition and the appearance of heavy quasiparticles with large effective masses. The first and most popular of these approaches is the slave-boson method, which we briefly describe here to give a general perspective on this class of methods.

Within the slave-boson approach the physical fermion, which is associated with the annihilation operators $c_{i\sigma}$ (here we limit to a single band model for simplicity) is represented in terms of a pseudofermion variable $f_{i\sigma}$ which has fermionic statistics and a boson variable $b_{i\sigma}$, where here the spin index of the boson operator does refer to an internal spin degree of freedom, but it should be viewed as a label which establishes the link between the slave boson and the correspondent fermion.

$$c_{i\sigma} \rightarrow f_{i\sigma} b_{i\sigma} \tag{4.30}$$

In this way the original Hilbert space is enlarged to include also the auxiliary bosonic degree of freedom. The nickname of "slave" particles derives from a constraint which will impose that the auxiliary variable contains the same physical information of the fermionic variable. This would obviously be redundant if the model was to be solved exactly. The idea is instead to perform a suitable mean-field after which the pseudofermion variable describes the itinerant quasiparticle fraction of the electron, while the auxiliary boson describes its localized fraction.

We give some more details of one popular slave-boson approach introduced by Kotliar and Ruckenstein [20], where the slave bosons are introduced as a sort of label attached to each local electronic configuration.

Explicitly, on each lattice site one can have four physical states, the empty state $|0\rangle$, the states where there is one electron with a given spin $c^\dagger_{i\uparrow}|0\rangle \equiv |\uparrow\rangle$ and $c^\dagger_{i\downarrow}|0\rangle \equiv |\downarrow\rangle$ and the doubly occupied configuration $c^\dagger_{i\uparrow}c^\dagger_{i\downarrow}|0\rangle \equiv |\uparrow\downarrow\rangle$. On each site we introduce four bosons which "label" these four states, $e^\dagger$, $s^\dagger_\uparrow$, $s^\dagger_\downarrow$, $d^\dagger$. When the fermions define a given local configuration, the bosons have to follow and the boson with the correct "label" must be created. The original Hilbert space can be mapped onto an extended space of bosons and fermions according to the following correspondence (repeated for every lattice site)

$$|0\rangle \Longleftrightarrow e^\dagger|\tilde{0}\rangle \tag{4.31}$$

$$c^\dagger_\uparrow|0\rangle = |\uparrow\rangle \Longleftrightarrow f^\dagger_\uparrow s^\dagger_\uparrow|\tilde{0}\rangle \tag{4.32}$$

$$c^\dagger_\downarrow|0\rangle = |\downarrow\rangle \Longleftrightarrow f^\dagger_\downarrow s^\dagger_\downarrow|\tilde{0}\rangle \tag{4.33}$$

$$c^\dagger_\uparrow c^\dagger_\downarrow|0\rangle = |\uparrow\downarrow\rangle \Longleftrightarrow f^\dagger_\uparrow f^\dagger_\downarrow d^\dagger|\tilde{0}\rangle \tag{4.34}$$

In the new Hilbert space whenever the fermionic degrees of freedom describe a given local configuration, the bosons have to follow, which means that one (and only one) given boson must exist on every lattice site. This constrained is enforced through a Lagrange multiplier which adds to the constraint which imposes that the physical state on each lattice site is equivalently described by fermionic and bosonic variables. One easily realizes that the creation operator for a physical fermion can be represented as

$$c_\sigma^\dagger \Longleftrightarrow f_{i\sigma}^\dagger (s_{i\sigma}^\dagger e_i + d^\dagger s_{i\bar{\sigma}}) o_i, \tag{4.35}$$

where o_i is any bosonic operator which gives 1 if evaluated on the physical Hilbert space. Clearly, if the original problem could not be solved exactly, there is no hope to solve the new problem, which is severely more complicated, as it now features a fermion-boson interaction and several constraints. The idea behind the whole construction is to perform a rather crude mean-field approximation which in turn leads to a non-interacting effective fermionic problem in which the physics of strong correlation is included in mean-field parameters arising from the bosonic degrees of freedom.

In the bosonic case, the way the mean-field is formulated is by taking a saddle-point approximation for all the bosonic degrees of freedom. As a matter of fact this amounts to replace every bosonic field in a path integral representation with a c-number. This can be pictured as a Bose condensation of the slave bosons. A crucial point is that one can easily see that the original fermionic interaction can be expressed exclusively in terms of bosons, which means that, in the saddle-point approximation it will become a number and the remaining fermionic Hamiltonian will be non-interacting and therefore easily solvable.

We do not enter here in the details of the slave-boson method, but we emphasize that the main result is that the interacting system will be described as an effective non-interacting system with a renormalization of the kinetic energy which essentially entails all the effects of the interactions. The renormalization is a momentum-independent number, which indeed coincides with the quasiparticle weight Z.

The correlated metal is therefore described as a Fermi liquid as long as Z is finite, and the Mott–Hubbard metal-insulator transition is associated with a vanishing of Z, which corresponds, within the slave-boson mean-field, to a divergent effective mass.

The approach followed in the slave-spin method, that we present in the next section, is analogous in many ways to the slave-boson mean-field. The respective physical descriptions of the modeled system share many of their main features.

4.5 The Slave-Spin Formalism and Its Mean-Field

In this section we introduce in some details the slave-spin mean-field method. This approach represents one of the simplest and computationally cheap approaches to study the strong correlation physics arising from large short-ranged interactions. As we anticipated above, this method belongs to the broader field of slave-particle methods, in which auxiliary particles are introduces to represent local degrees of freedom in a strongly correlated system. In the previous section we have sketched the most popular of these approaches, the slave bosons, where local bosons are introduced to represent local electronic configurations. While this method is appealing because one can easily perform a saddle-point approximation of the bosonic degrees of freedom, it is evident that the use of bosonic degrees of freedom to represent the same physics of fermions is hugely redundant. Bosons, with an infinite Hilbert space, are in fact used to describe the same physics as fermions whose Hilbert space comprises only two states. In the Kotliar–Ruckenstein approach [20] this implies two constraints that impose that one and only one boson exists at every site, and that its "label" corresponds to the fermionic state.

Within the slave-spin approach, we reduce the redundancy and get rid of one of the two constraints by introducing spin-1/2 auxiliary variables in correspondence with each fermionic operator.

In this section we introduce the slave-spin approach and its mean field in the version of [21, 22]. Another version of the method, completely equivalent for the purpose of this section has been proposed in [23] and applied on the Fe-superconductors (see references in Sect. 4.7). Further studies on interesting different developments of this technique are [24–28]. As we will see the slave-spin mean field provides us with a simple and fast method to study strongly correlated metals and their disappearance into a Mott transition. Just like in the slave-boson method, the metallic solution will be described as Fermi liquids all the way to the Mott transition. One important aspect of the slave-spin method is that it can be easily generalized to multi-orbital models, as we will discuss in Sect. 4.6, where it remains numerically cheap and agile compared to similar methods.

In the slave-spin representation, we map the original local Fock space of the problem onto a larger local Fock space that contains as many fermionic degrees of freedom (named $f_{i\sigma}$) as the original plus the same number of spin-1/2 quantum variables, one for each $f_{i\sigma}$. It is worth noting that the auxiliary spins have nothing to do with the physical spin of the electrons, but they are just auxiliary variables which have been chosen to have the commutation relations of spins. In fact a slave-spin variable is introduced for every fermion species, taking into account the fermion spin multiplicity, so that slave-spins are also labeled with a physical-spin index σ i.e., $S^z_{i\sigma}$.

We then associate to every state of the original physical space one of the states in this larger space by using the correspondence:

$$|n^d_{i\sigma} = 1\rangle \Longleftrightarrow |n^f_{i\sigma} = 1,\ S^z_{i\sigma} = +1/2\rangle, \tag{4.36}$$

$$|n^d_{i\sigma} = 0\rangle \Longleftrightarrow |n^f_{i\sigma} = 0,\ S^z_{i\sigma} = -1/2\rangle. \tag{4.37}$$

In words, when a local orbital and spin state is occupied then the corresponding slave-spin is "up" and if it is empty the slave-spin is "down". With these one-particle local states one construct the many-particle states as usual.

The enlarged local Fock space contains also unphysical states such as $|n^f_{i\sigma} = 0, S^z_{i\sigma} = +1/2\rangle$ and $|n^f_{i\sigma} = 1, S^z_{i\sigma} = -1/2\rangle$. These unphysical states are excluded if the following local constraint is enforced at each site and for each σ:

$$f^\dagger_{i\sigma} f_{i\sigma} = S^z_{i\sigma} + \frac{1}{2}. \tag{4.38}$$

The next step is to map the physical operators onto operators that act in the enlarged Fock space. The electron number operator is equivalently represented by the auxiliary fermions number, i.e., $n^d_{i\sigma} \rightarrow n^f_{i\sigma}$, but also by the z component of the slave-spin $n^d_{i\sigma} \rightarrow S^z_{i\sigma} + 1/2$. This allows us to rewrite any density-density interaction term in the hamiltonian in terms of the spins only. This is a crucial point, which recalls the fact that in the slave-boson language the interaction can be recast in terms of bosons only and will be crucial to eventually obtain an effective non-interacting Hamiltonian for the fermionic variables.

For the single-band Hubbard model it is convenient to use the particle-hole symmetric form (4.29):

$$H_{int}[d] = \frac{U}{2}\sum_i \left(\sum_\sigma \left(n^d_{i\sigma} - \frac{1}{2}\right)\right)^2 \longrightarrow H_{int}[S] = \frac{U}{2}\sum_i \left(\sum_\sigma S^z_{i\sigma}\right)^2 \tag{4.39}$$

For the non-diagonal operators appearing in the hopping terms we have some freedom of choice, equivalent to the choice of the operators o_i in (4.35). Indeed several operators will have the same action on the physical states while differing when acting on the non-physical ones. Any of these operators is completely equivalent, as long as the constraint (4.38) is enforced exactly, excluding non-physical states. For example both the choices $f^\dagger_{i\sigma} S^+_{i\sigma}$ and $f^\dagger_{i\sigma} 2S^x_{i\sigma}$ will have the same action on the physical state $|n^f_{i\sigma} = 0,\ S^z_{i\sigma} = -1/2\rangle$, transforming it into $|n^f_{i\sigma} = 1,\ S^z_{i\sigma} = +1/2\rangle$, like does $d^\dagger_{i\sigma}$ on the corresponding original state, i.e., $d^\dagger_{i\sigma}|n^d_{i\sigma} = 0\rangle = |n^d_{i\sigma} = 1\rangle$.

In the following we derive the most general form of the spin-part of the composite operator. Indeed we can write a generic expression

$$d_{i\sigma} \rightarrow f_{i\sigma} O_{i\sigma}, \qquad d^\dagger_{i\sigma} \rightarrow f^\dagger_{i\sigma} O^\dagger_{i\sigma} \tag{4.40}$$

in which $O_{i\sigma}$ is a generic spin-1/2 operator, i.e., a 2×2 complex matrix. It is easy to determine that the most general form for $O_{i\sigma}$ is

$$O_{i\sigma} = \begin{pmatrix} 0 & c_{i\sigma} \\ 1 & 0 \end{pmatrix} = S^-_{i\sigma} + c_{i\sigma} S^+_{i\sigma}, \tag{4.41}$$

where $c_{i\sigma}$ is an arbitrary complex number.

Let's check that the operators (4.40) have, in the physical states of the enlarged Fock space, the same effect as the fermionic operators in the original Fock space, i.e.,

$$\begin{aligned} d_{i\sigma}|n^d_{i\sigma} = 0\rangle &= 0, \quad & d_{i\sigma}|n^d_{i\sigma} = 1\rangle &= |n^d_{i\sigma} = 0\rangle \\ d^\dagger_{i\sigma}|n^d_{i\sigma} = 1\rangle &= 0, \quad & d^\dagger_{i\sigma}|n^d_{i\sigma} = 0\rangle &= |n^d_{i\sigma} = 1\rangle \end{aligned} \tag{4.42}$$

The two conditions on the right determine three out of four elements of $O_{i\sigma}$.

$$f^\dagger_{i\sigma} O^\dagger_{i\sigma} | n^f_{i\sigma}{=}0, S^z_{i\sigma}{=}{-1/2}\rangle = | n^f_{i\sigma}{=}1, S^z_{i\sigma}{=}{+1/2}\rangle, \quad (4.43)$$

$$f_{i\sigma} O_{i\sigma} | n^f_{i\sigma}{=}1, S^z_{i\sigma}{=}{+1/2}\rangle = | n^f_{i\sigma}{=}0, S^z_{i\sigma}{=}{-1/2}\rangle, \quad (4.44)$$

imply

$$O^\dagger_{i\sigma} | S^z_{i\sigma} = -1/2\rangle = | S^z_{i\sigma} = +1/2\rangle, \quad (4.45)$$

$$O_{i\sigma} | S^z_{i\sigma} = +1/2\rangle = | S^z_{i\sigma} = -1/2\rangle, \quad (4.46)$$

which impose $O_{i\sigma;11} = 0$, $O_{i\sigma;21} = 1$, and $O^*_{i\sigma;22} = 0$.

The two conditions on the left hand side of (4.42) are instead assured by the fermionic operators $f_{i\sigma}$, i.e.,

$$f_{i\sigma} O_{i\sigma} | n^f_{i\sigma} = 0, S^z_{i\sigma} = -1/2\rangle = 0, \quad (4.47)$$

$$f^\dagger_{i\sigma} O^\dagger_{i\sigma} | n^f_{i\sigma} = 1, S^z_{i\sigma} = +1/2\rangle = 0 \quad (4.48)$$

for any $O_{i\sigma}$. This implies that $O_{i\sigma;12} = c_{i\sigma}$ remains undetermined.[6]

The arbitrariness of the complex number $c_{i\sigma}$ is a sort of gauge of our formulation as long as the constraint is treated exactly. In any practical implementation some approximation has to be performed to enforce the constraint in our mean-field treatment. In these approximations the particular choice of the gauge comes into play. $c_{i\sigma}$ can indeed be tuned in order to give rise to the most physical approximation scheme, by imposing, for instance, that the mean field correctly reproduces solvable limits of the problem, like the non-interacting limit. A similar freedom is indeed present also in the Kotliar–Ruckenstein slave-boson mean-field, where the choice of reproducing the non-interacting limit implies that the mean-field coincides with the Gutzwiller approximation [20].

We will see that the correct choice of $c_{i\sigma}$ depends on the average occupation of the local state, and is such that it reduces to 1 at occupation $1/2$, so that $O_{i\sigma} = 2S^x_{i\sigma}$ and the prescription for half-filling is recovered.

Thus finally, in the enlarged Fock space the Hamiltonian can be written exactly as:

$$\hat{H} - \mu\hat{N} = \sum_{<ij>\sigma} t_{ij} O^\dagger_{i\sigma} O_{j\sigma} f^\dagger_{i\sigma} f_{j\sigma} + \frac{U}{2} \sum_i \left(\sum_\sigma S^z_{i\sigma} \right)^2 - \mu \sum_{i\sigma} n^f_{i\sigma}, \quad (4.49)$$

subject to the constraint (4.38).

[6]For any choice of $c_{i\sigma}$ the operators (4.40) duly respect anticommutation relations, in the subspace of physical states.

4.5.1 Mean-Field Approximation

So far we have only formulated a more involuted and redundant representation for the original Hubbard model, which obviously does not allow us to solve exactly the model. In the same spirit of the slave-boson mean field, we now perform a mean-field approximation, which consists of three main steps

1. Decoupling auxiliary fermions and slave-spin degrees of freedom
2. Treating the constraint on average, using a site- and spin-independent (since we will investigate uniform non-magnetic phases) Lagrange multiplier λ
3. Solving the resulting slave-spin Hamiltonian in a single-site mean-field approximation (a simple Weiss-like mean-field).

As we will see, this approximation leads to an effective non-interacting Hamiltonian for the pseudofermions similar to the result of the slave-boson mean-field. In the present case, the parameters of the pseudofermion Hamiltonian are determined from the solution of the slave-spin Hamiltonian rather than from the saddle-point approximation for the slave bosons. Let us now discuss the mean-field solution of the Hubbard model in details.

4.5.1.1 Decoupling Fermions from Spins

The main obstacle to solve the new Hamiltonian (4.49) lies in the original hopping terms, which has now turned into an interaction between pseudofermions and spins. We thus decouple the mixed spin-fermion hopping term according to

$$\sum_{<ij>\sigma} t_{ij} O^{\dagger}_{i\sigma} O_{j\sigma} f^{\dagger}_{i\sigma} f_{j\sigma} \simeq \sum_{<ij>\sigma} t_{ij} \langle O^{\dagger}_{i\sigma} O_{j\sigma} \rangle f^{\dagger}_{i\sigma} f_{j\sigma} + \sum_{<ij>\sigma} t_{ij} O^{\dagger}_{i\sigma} O_{j\sigma} \langle f^{\dagger}_{i\sigma} f_{j\sigma} \rangle . \tag{4.50}$$

As a result the Hamiltonian can be rewritten as a sum of a purely fermionic one and a purely spin one: $H = H_f + H_S$. The effect of the original interaction now lies in the fact that the coefficients of the two Hamiltonians depend on expectation values computed on the other.

4.5.1.2 Treating the Constraint on Average

Despite the separation of Hamiltonians, the problem remains practically intractable because of the constraint which still couples the two degrees of freedom. We can write the partition function in the enlarged Fock space as:

$$\mathcal{Z} = Tr\left[e^{-\beta(\hat{H}-\mu\hat{N})} \delta\left(S^z_{i\sigma} + \frac{1}{2} - f^{\dagger}_{i\sigma} f_{i\sigma} \right) \right] \tag{4.51}$$

where the delta function imposes the constraint by projecting out all contributions from the unphysical states. Even having decoupled the hamiltonian in step 1, because of the delta function, the sum over the enlarged Fock space cannot be replaced by two factorized sums over the decoupled spaces of spin and fermions.

But one can relax the constraint by replacing the delta inside the trace with a simple exponential of the form $e^{-\beta\lambda(S^z_{i\sigma}+\frac{1}{2}-f^\dagger_{i\sigma}f_{i\sigma})}$, and realize that λ can be adjusted so that the constraint is respected on average, i.e., it is satisfied by the expectation values of the operators. Indeed by minimizing the grand potential $\Omega = -K_B T \log \mathcal{Z}$ with respect to λ, that is determining what λ satisfies the condition $\partial\Omega/\partial\lambda = \frac{1}{\mathcal{Z}}Tr[e^{-\beta(\hat{H}-\mu\hat{N})}(S^z_{i\sigma}+\frac{1}{2}-f^\dagger_{i\sigma}f_{i\sigma})] = 0$, one ensures that $\langle S^z_{i\sigma}\rangle + \frac{1}{2} = \langle f^\dagger_{i\sigma}f_{i\sigma}\rangle$. Thus λ acts as a Lagrange multiplier for the constrained minimization of Ω by respect to the mean-field parameters.

After the first two steps, the total Hamiltonian can be written as the sum of the following two effective Hamiltonians:

$$H_f = \sum_{<ij>,\sigma} t_{ij}(Q_{ij} f^\dagger_{i\sigma} f_{j\sigma} + H.c.) - (\mu+\lambda)\sum_i n^f_i, \tag{4.52}$$

$$H_s = \sum_{<ij>,\sigma} (J_{ij} O^\dagger_{i\sigma} O_{j\sigma} + H.c.) + \lambda \sum_{i,\sigma}(S^z_{i\sigma} + \frac{1}{2}) + \frac{U}{2}\sum_i \left(\sum_\sigma S^z_{i\sigma}\right)^2 . \tag{4.53}$$

The parameters Q_{ij} (hopping renormalization factor), J_{ij} (slave-spin exchange constant) and λ_i in these expression are determined from the following coupled self-consistency equations:

$$Q_{ij} = \langle O^\dagger_{i\sigma} O_{j\sigma}\rangle_s, \tag{4.54}$$

$$J_{ij} = t_{ij}\langle f^\dagger_{i\sigma} f_{j\sigma}\rangle_f, \tag{4.55}$$

$$\langle n^f_{i\sigma}\rangle_f = \langle S^z_{i\sigma}\rangle_s + \frac{1}{2}, \tag{4.56}$$

where $\langle\rangle_{f,s}$ indicates the effective Hamiltonian and the corresponding Fock space used for the calculation of the averages.

4.5.1.3 Mean-Field Approximation for the Slave-Spin Hamiltonian

We are thus left with two Hamiltonians, one for the pseudofermions and the other for the slave spins, which must be solved in a self-consistent way so that the conditions (4.54)–(4.56) are fulfilled.

The pseudofermionic Hamiltonian is indeed a renormalized independent-fermion Hamiltonian for the $f_{i\sigma}$ of which the hopping is renormalized by the expectation values of some spin operators. However, the spin Hamiltonian retains essentially the full complexity of the original model. We have thus to perform a further approximation, in this case the single-site mean-field, on the spin Hamiltonian.

For this, a Hamiltonian with one site only is considered, on which the interaction is treated exactly, and is embedded in the effective ("Weiss") field of its surroundings, i.e., the quantum and thermodynamic average value of the operators on other sites coupled to the chosen site. Assuming translational invariance this approximate Hamiltonian is self-consistently used to calculate also the same mean-field average values.

Mathematically this means performing the approximation

$$\langle O^\dagger_{i\sigma} O_{j\sigma}\rangle \simeq \langle O^\dagger_{i\sigma}\rangle O_{j\sigma} + O^\dagger_{i\sigma}\langle O_{j\sigma}\rangle \tag{4.57}$$

so that we obtain the following mean-field slave-spin Hamiltonian:

$$H_s = \sum_i H^i_s = \sum_{i,\sigma}(h_{i\sigma} O^\dagger_{i\sigma} + H.c.) + \lambda \sum_{i,\sigma}\left(S^z_{i\sigma} + \frac{1}{2}\right) + \frac{U}{2}\sum_i\left(\sum_\sigma S^z_{i\sigma}\right)^2. \tag{4.58}$$

where

$$h_{i\sigma} = h_\sigma = \sum_j J_{ij}\langle O_{j\sigma}\rangle_s = \langle O_{i\sigma}\rangle_s \sum_j t_{ij}\langle f^\dagger_{i\sigma} f_{j\sigma}\rangle_f, \tag{4.59}$$

given the definition of J_{ij} and the fact that $\langle O_{i\sigma}\rangle = \langle O_{j\sigma}\rangle$ because of translational invariance. This last sum is independent on the renormalized parameters in the one-band case and it can be evaluated once and for all, yielding:

$$\sum_j t_{ij}\langle f^\dagger_{i\sigma} f_{j\sigma}\rangle_f = \sum_{\mathbf{k}} \epsilon_{\mathbf{k}}\langle f^\dagger_{\mathbf{k}\sigma} f_{\mathbf{k}\sigma}\rangle_f \equiv \bar{\epsilon}, \tag{4.60}$$

which is the kinetic energy per spin for the pseudofermions.

Equations (4.52) and (4.58), with the definitions (4.54) and (4.59), and the implicit condition (4.56) constitute the slave-spin mean-field equations for the solution of the single-band Hubbard model (4.23), and have to be solved for the mean-field parameters $h_{i\sigma}$, Q_{ij}, λ. It is worth noting that the hopping renormalization factor Q_{ij}, after the last approximation becomes $Q_{ij} = \langle O_{i\sigma}\rangle^2$ and it is the same for every pair of sites. As a result it becomes a prefactor which multiplies a kinetic energy operator identical to the non-interacting result. The physical picture emerging is that of a kinetic energy renormalized by a factor, which can be shown to coincide with the quasiparticle weight Z, as we show explicitly in the complement Sect. 4.5.3. This is a common result to the single-site mean-field of slave particle methods and it mirrors the natural expectation that the motion of electrons in the presence of strong interaction is reduced.

Turning back to our solution, the single-site approximation also implies that the slave-spin hamiltonian is the sum of decoupled one-site hamiltonians so that $h_{i\sigma} = h_\sigma$ in (4.59) and the sum over i can be dropped in (4.58). The practical advantage is now that the slave-spin one-site hamiltonian can be solved exactly as it has a Fock space of finite dimension. Notice that including more orbitals in the modelling will increase the size of this local Hilbert space introducing one spin for every fermionic variable. The growth of this local Hilbert space will be the bottleneck of actual calculations.

In practice these equations are solved iteratively: starting with a guess for Z and λ one solves the fermionic hamiltonian in k-space and calculates $h_{i\sigma}$, which is used in the spin hamiltonian to calculate the new Z. Within the cycle λ is adjusted so that the average constraint (4.56) is satisfied, and the scheme is iterated until convergence of all quantities.

The only part of the scheme yet to be specified is the choice of the complex number $c_{i\sigma}$ ($c_{i\sigma} = c$ in the present site- and spin-independent mean-field) [22]. The most natural choice is to use this gauge of our formulation to impose that the non-interacting limit, which is exactly soluble, is exactly reproduced, i.e., that Z(U = 0) = 1. Imposing this condition (and taking c real) gives an expression for c that depends only on the density $\langle n_{d\sigma}\rangle = \langle n_{f\sigma}\rangle \equiv n_\sigma$, i.e.,

$$c = \frac{1}{\sqrt{n_\sigma(1-n_\sigma)}} - 1 \tag{4.61}$$

The gauge can be fixed then once and for all for a given wanted density, that is reached by adjusting the chemical potential μ. In practice it can be advantageous to adjust μ and c along with the other parameters during the mean-field iteration scheme just outlined.

With this choice we see from Fig. 4.2 that at the same fixed population n the single-site mean field of the Slave Spins method gives exactly the same results of the Kotliar–Ruckenstein mean-field of the Slave Bosons method, i.e., the Gutzwiller Approximation.

4.5.2 Complement: *Derivation of the Gauge c for Arbitrary Filling*

In the single-site approximation, we can determine the gauge c analytically [22, 29].

The non-interacting single-site slave spin Hamiltonian H_s reads:

$$\hat{H}_s = hO^\dagger + h^* O + \lambda S^z, \tag{4.62}$$

where O is defined as in (4.41). The single-site fermionic part of the Hamiltonian (where we have absorbed the $\lambda/2$) is simply spinless non-interacting fermions. The physical spin index σ is suppressed in H_s since for $U = 0$ upspin and downspin fermions are decoupled, so that we can diagonalize the hamiltonian for one slave-spin in the $S^z = \pm 1/2$ basis. Thus the matrix to be diagonalized reads:

$$H_s = \begin{pmatrix} \frac{\lambda}{2} & a \\ a^* & -\frac{\lambda}{2} \end{pmatrix} \tag{4.63}$$

and $a = h + ch^*$.

The ground state eigenvalue ϵ_{GS} and the corresponding eigenstate are

$$\epsilon_{GS} = -\sqrt{\frac{\lambda^2}{4} + |a^2|} \equiv -R \tag{4.64}$$

$$|GS\rangle = \begin{pmatrix} \frac{\frac{\lambda}{2}+R}{N} \\ \frac{-a*}{N} \end{pmatrix} \tag{4.65}$$

with $N = \sqrt{2R(\frac{\lambda}{2} + R)}$.

The expectation value of S^z and O in the ground state are

$$\langle S_z \rangle = \frac{\lambda}{4R} \tag{4.66}$$

and

$$\langle O \rangle = -\frac{ca^* + a}{2R} \tag{4.67}$$

The Lagrange multiplier depends on the density $n = \sum_\sigma n_\sigma$ and is adjusted in order to satisfy the constraint equation:

$$n_\sigma - \frac{1}{2} = \langle S_z \rangle = \frac{\lambda}{4R} \tag{4.68}$$

We want to tune c in order to match the condition that in the limit $U = 0$ the renormalization factor Z must be unity:

$$Z =< O >^2 = \frac{|ca^* + a|^2}{4R^2} = 1 \tag{4.69}$$

We can easily eliminate λ from these two conditions, by squaring (4.68). We are left with the following expression for c:

$$\frac{|a|^2}{|ca^* + a|^2} = \frac{1}{|ce^{-i \arg a} + e^{i \arg a}|} = n_\sigma - n_\sigma^2 \tag{4.70}$$

If we choose c to be real then h and a are also real. Then, the expression for c in the closed form is

$$c = \frac{1}{\sqrt{n_\sigma(1 - n_\sigma)}} - 1. \tag{4.71}$$

Note that this result is independent of h.

4.5.3 Complement: *Fermi-Liquid Quasiparticle Weight and Mass Enhancement*

Using the correspondence $d_{i\sigma} \to O_{i\sigma} f_{i\sigma}$ we can rewrite the physical Green function, within our mean-field formalism as a product of fermionic and spin Green's functions:

$$G^d_{ij}(t) = G^f_{ij}(t) G^S_{ij}(t) \tag{4.72}$$

with

$$G^d_{ij}(t) = -i \langle T_t d_{i\sigma}(t) d^\dagger_{j\sigma}(0) \rangle_d \tag{4.73}$$

$$G^f_{ij}(t) = -i \langle T_t f_{i\sigma}(t) f^\dagger_{j\sigma}(0) \rangle_f \tag{4.74}$$

$$G^S_{ij}(t) = \langle T_t O_{i\sigma}(t) O^\dagger_{j\sigma}(0) \rangle_S \tag{4.75}$$

a Fourier transform with respect to time transforms the equal time product into a convolution

$$G^d_{\mathbf{k}}(\omega) = \int \frac{d\omega'}{2\pi} \frac{d\mathbf{k}'}{8\pi^3} G^f_{\mathbf{k}'}(\omega') G^S_{\mathbf{k}-\mathbf{k}'}(\omega - \omega'). \tag{4.76}$$

In the metallic phase the slave-spins are long-range ordered, this implies that the correlation function G^S has a finite value at large distances and times which is given by $Z = \langle O^\dagger_{i\sigma} \rangle \langle O_{j\sigma} \rangle$, and thus its Fourier transform has a Dirac delta contribution at zero frequency

$$G^S_{\mathbf{k}}(\omega) = Z \delta(\omega) \delta(\mathbf{k}) + G^{S,reg}_{\mathbf{k}}(\omega) \tag{4.77}$$

which in turn implies that

$$G^d_{\mathbf{k}}(\omega) = Z G^f_{\mathbf{k}}(\omega) + \int \frac{d\omega'}{2\pi} \frac{d\mathbf{k}'}{8\pi^3} G^f_{\mathbf{k}'}(\omega') G^{S,reg}_{\mathbf{k}-\mathbf{k}'}(\omega - \omega'). \tag{4.78}$$

It is thus obvious that Z plays the role of a quasiparticle weight.

When comparing the expressions of the particle ($d^\dagger_{i\sigma}$) and quasiparticle ($f^\dagger_{i\sigma}$) Green functions, i.e.,:

$$G^d_{\mathbf{k}}(\omega) = \frac{1}{\omega + \mu - \epsilon_{\mathbf{k}} - \Sigma(\omega)} \tag{4.79}$$

and

$$G^f_{\mathbf{k}}(\omega) = \frac{1}{\omega + \mu + \lambda - Z\epsilon_{\mathbf{k}}} \tag{4.80}$$

one sees that the electronic self-energy is, in this scheme:

$$\Sigma(\omega) = \mu - \frac{\mu + \lambda}{Z} + \omega \left(1 - \frac{1}{Z}\right). \tag{4.81}$$

This establishes that the slave-spin mean-field, in complete analogy with the slave-boson mean-field and the Gutzwiller approximation, is equivalent to a self-energy which features a low-frequency linear behavior which determines the quasiparticle renormalization and a constant term, which determines a shift of the chemical potential.

Before entering in the main argument of this chapter, which is the application of the slave-spin mean-field to the multi orbital models for iron-based superconductors, we present some benchmark calculations to assess the ability of the approach to study well-known limits and regimes.

4.5.4 The Half-Filled Single-Band Hubbard Model

We begin our exploration of simple results within SSMF with the half-filled single-band Hubbard model. In Fig. 4.1 we report the quasiparticle weight obtained by solving the slave-spin mean-field equations for a half-filled system with a semicircular density of states with semi-bandwidth D. In this case we can obtain $\lambda = 0$ from particle-hole symmetry and solve the remaining equations numerically. The results share the same qualitative and quantitative behavior of popular mean-field approaches to the Mott transition, like the slave-boson mean-field (Brinkmann–Rice transition) and the Gutzwiller approximation. Z decreases regularly from the non-interacting value 1 and vanishes at a critical value $U_c \simeq 3.39D$ (see complement Sect. 4.5.5 for an analytical derivation of this value), slightly larger than the result obtained by means of Dynamical Mean-Field Theory, which captures the local quantum fluctuations beyond mean-field. We recover that a vanishing Z amounts to a divergent effective mass. The main problem of this approach is that it is limited to the strongly correlated metallic region and its description of the Mott insulating state is basically that of the atomic limit.

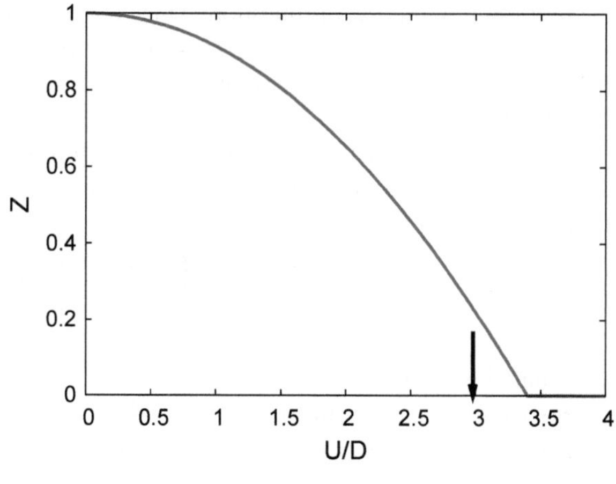

Fig. 4.1 Quasiparticle weight in the half-filled single-band Hubbard model solved with SSMF, for a semi-circular DOS of half-bandwidth D. The *arrow* show the U_c obtained in DMFT for the same model

4.5.5 Complement: *Critical U for the Mott Transition in the Single-Orbital Hubbard Model*

Here we show how to obtain analytically within the slave-spin mean-field approximation the critical interaction strength U_c for the Mott transition in the half-filled one-band Hubbard model.

Indeed the slave-spin equation (4.58) can be developed around the insulating solution $h_{i\sigma} = 0$ where $H_s = H_{at}$ with

$$H_{at} = \frac{U}{2}\left(\sum_\sigma S^z_\sigma\right)^2, \tag{4.82}$$

of with the spectrum can be calculated exactly. One can calculate the effect, as a perturbation, of the "kinetic" part of the slave-spin hamiltonian:

$$H_{pert} = h \sum_\sigma 2S^x_\sigma. \tag{4.83}$$

where $O_{i\sigma} = 2S^x_\sigma$ because for a half-filled band the gauge $c = 1$ ((4.61) for n = 1/2), and we call $h \equiv h^*_{i\sigma} + h_{i\sigma}$. Also $\lambda = 0$ at half-filling because of particle-hole symmetry.

The perturbative parameter here is thus h, which is small near the metal-insulator transition, if this is second-order.

The unperturbed space is:

$$\begin{cases} |\uparrow\uparrow\rangle \; E = \frac{U}{2} \\ |\uparrow\downarrow\rangle \; E = 0 \\ |\downarrow\uparrow\rangle \; E = 0 \\ |\downarrow\downarrow\rangle \; E = \frac{U}{2} \end{cases} \tag{4.84}$$

The unperturbed ground state being two times degenerate one has to use the degenerate perturbation theory. Furthermore, since the perturbation H_{pert} does not have any nonzero elements in the low-energy subspace one has to use second-order perturbation theory to find the state, in the ground-state subspace, to which the perturbed state tends for $h \to 0$.

In second-order perturbation one has to diagonalize the matrix $H' \equiv H_{pert}(E_0 - H_{at})^{-1} H_{pert}$ in the degenerate subspace to obtain the energy corrections at the leading order. In this case:

$$H' = \begin{pmatrix} -\frac{4h^2}{U} & -\frac{4h^2}{U} \\ -\frac{4h^2}{U} & -\frac{4h^2}{U} \end{pmatrix} \tag{4.85}$$

diagonalising which one obtains $|\phi_0\rangle \equiv (|\uparrow\downarrow\rangle + |\downarrow\uparrow\rangle)/\sqrt{2}$ as the ground state, of energy (to second order in h) $E_0^{(II)} = -\frac{8h^2}{U}$.

The correction to the ket is however nonzero at first order in h and reads:

$$|\phi_0^{(I)}\rangle = |\phi_0\rangle - \frac{2}{U}2h(|\uparrow\uparrow\rangle + |\downarrow\downarrow\rangle)/\sqrt{2}. \tag{4.86}$$

To calculate the critical interaction one simply uses the self-consistency condition $h = 2\bar{\epsilon}\langle 2S_\sigma^x\rangle$. This condition has to be rigorously satisfied and since we can only calculate here a perturbative development for $\langle 2S_\sigma^x\rangle$, which by definition is only exactly valid at the transition, (where $h \to 0$), the self-consistency condition becomes an equation determining the coupling U for which this condition can be satisfied rigorously, hence the critical coupling U_c.

The calculation for $\langle 2S_\sigma^x\rangle$ on the state $|\phi_0^{(I)}\rangle$ gives:

$$\langle\phi_0^{(I)}|2S_\sigma^x|\phi_0^{(I)}\rangle = -\frac{8h}{U}. \tag{4.87}$$

The self-consistency equation becomes then:

$$h = 2\bar{\epsilon}\langle 2S_\sigma^x\rangle = -\frac{16h}{U_c}\bar{\epsilon} \implies U_c = -16\bar{\epsilon}. \tag{4.88}$$

For a semi-circular DOS of half-bandwidth $D = 1$, $\bar{\epsilon} \simeq -0.2122$, which gives

$$U_c \simeq 3.39 \tag{4.89}$$

if $D = 1$, in agreement with the numerical result Fig. 4.1.

4.5.6 The Hubbard Model for Finite Doping

The treatment of the Hubbard within SSMF model can be easily extended to any density away from half-filling. This is indeed obtained by moving the chemical potential away from its particle-hole symmetric value μ ($= 0$ or $= U/2$ depending on which among the equivalent forms of the Hubbard hamiltonian are used - see complement Sect. 4.3.1 - these analytically-derived values hold for particle-hole symmetric DOS), and determining the corresponding value of the Lagrange multiplier $\lambda \neq 0$ at the relevant doping. As mentioned in Sect. 4.5.2 the gauge (4.61) in the slave-spin formalism depends on the final population, so that its value has to be updated in the numerical solution of the mean-field equations.

It can be shown (Fig. 4.2) that this formulation gives exactly the same results than the Kotliar–Ruckenstein slave-boson mean-field [20], as illustrated in Fig. 4.2. The quasiparticle weight is still reduced with increasing interaction but does not vanish. Indeed the Mott transition needs an integer number of electron per site. For large values of the doping away from half-filling it tends to the proper value $Z = 1$ for a dilute electron system, while at small doping indeed Z is strongly reduced following the trend of the half-filled case, but then it saturates at a small finite value that at large

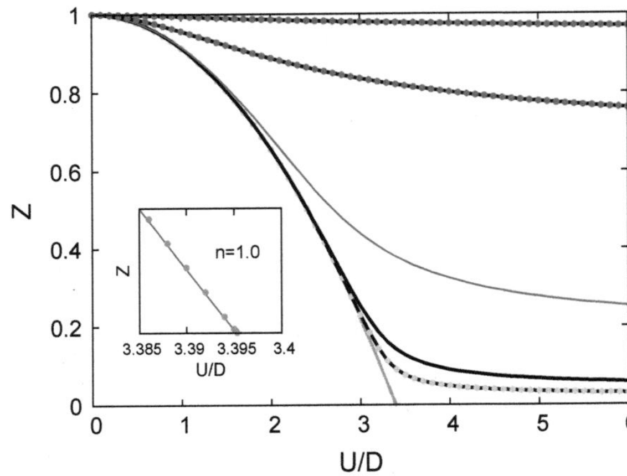

Fig. 4.2 Comparison of Slave-Spin (within the gauge (4.61), *continuous lines*) and Kotliar–Ruckenstein Slave-bosons (*dots*) single-site mean-fields in the one-band Hubbard model. For any fixed population (from below, the filling is $n = \sum_\sigma n_\sigma = 1.0, 0.99, 0.98, 0.9, 0.5, 0.1$) the two methods give coincident results. *Inset* blow-up of the Mott transition at half-filling, $n = 1.0$. From [22]

U depends only on the doping. For $U \rightarrow \infty$ the value of Z can be derived analytically [22] and it reproduces the value from the Gutzwiller approximation (equivalent to Kotliar–Ruckenstein slave-boson mean-field [20])

$$Z = \frac{2x}{1+x} \tag{4.90}$$

where $x = |1-n|$ is the absolute value of the doping (with $n = \sum_\sigma n_\sigma$ the total density).

4.6 Multi-orbital Correlations: Hund's Coupling

In this section we introduce one of the main characters which enter into play when the low-energy description of the material requires to include more than one orbital per site. So far we have considered the simplified case when the low energy model is restricted to one orbital only. In this situation there is obviously only one matrix element corresponding to the local interaction, i.e., the element with $m = m' = n = n'$ in (4.22), that we labeled U.

However when multiple orbitals are retained in the basis to expand the electron field, other integrals appear. The number of independent integrals depends on the number of retained orbitals and on the spatial symmetry of the problem. As discussed in Sect. 4.3.2 materials where the valence bands arise from 3d orbitals are the reference case for strong correlations. In these materials the five orbitals of the 3d shells are typically lifted in energy by a crystal field which has quite often cubic symmetry and separates a pair of e_g orbitals, from a triplet of t_{2g} orbitals. In these cases one can show that there are 3 independent integrals[7]:

$$
\begin{aligned}
V_{iiii}^{mmmm} &\equiv U = \int d\mathbf{r}d\mathbf{r}' \, |w_m(\mathbf{r})|^2 \, W(\mathbf{r},\mathbf{r}') \, |w_m(\mathbf{r}')|^2 \\
V_{iiii}^{mm'm'm} &\equiv U' = \int d\mathbf{r}d\mathbf{r}' \, |w_m(\mathbf{r})|^2 \, W(\mathbf{r},\mathbf{r}') \, |w_{m'}(\mathbf{r}')|^2 \\
V_{iiii}^{mm'mm'} &\equiv J = \int d\mathbf{r}d\mathbf{r}' \, w_m^*(\mathbf{r}) w_{m'}^*(\mathbf{r}') \, W(\mathbf{r},\mathbf{r}') \, w_m(\mathbf{r}') w_{m'}(\mathbf{r})
\end{aligned}
\tag{4.91}
$$

and all others vanish by symmetry. Thus the interaction used in the multi-orbital Hubbard model in these cases has customarily the form of the "Kanamori" hamiltonian[8]:

$$
\begin{aligned}
\hat{H}_{int} = U \sum_m n_{m\uparrow} n_{m\downarrow} &+ U' \sum_{m\neq m'} n_{m\uparrow} n_{m'\downarrow} + (U' - J) \sum_{m<m',\sigma} n_{m\sigma} n_{m'\sigma} + \\
&- J \sum_{m\neq m'} d^+_{m\uparrow} d_{m\downarrow} \, d^+_{m'\downarrow} d_{m'\uparrow} + J \sum_{m\neq m'} d^+_{m\uparrow} d^+_{m\downarrow} \, d_{m'\downarrow} d_{m'\uparrow}
\end{aligned}
\tag{4.92}
$$

Cubic symmetry already, for the e_g case, and the requirement of full "rotational" symmetry (i.e., invariance of the hamiltonian under orbital, spin and charge gauge symmetry transformations, separately) for the t_{2g} case impose the further condition

$$
U' = U - 2J. \tag{4.93}
$$

This form of the interaction Hamiltonian is not exact in symmetries other than cubic or if one considers more orbitals (as it is appropriate for a full 3d shell, for example), and anyway implies the rotational invariance of $W(r - r')$ which is true in free space but only approximate in a solid state environment. It is however customarily used in practice.

It is quite intuitive that it enforces the fact, transparent from the density-density terms, that the Coulomb repulsion between two electrons is reduced when these occupy two different orbitals. Moreover in this case the Pauli principle does not prevent the electrons to have their spins to be either anti-parallel or parallel along the quantization axis (say $\hat{z}$). In this case the purely quantum "exchange" integral J further lowers the energy of the parallel configuration. The spin-flip terms account for the same effect along the $\hat{x}$ and $\hat{y}$ directions, restoring the spin rotational invariance.

[7] All Wannier functions are intended here centered around the site with $R_i = 0$.

[8] The coefficient of the last term, the "pair-hopping", is equal to J only if the basis functions are chosen real-valued.

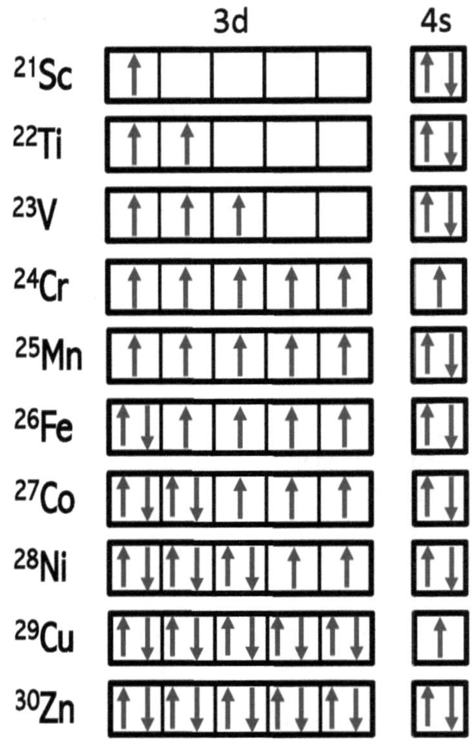

Fig. 4.3 Atomic Hund's rules in the 3d series of elements

In an isolated atoms these interaction terms originate the well-known "Hund's rules" that dictate the *aufbau* of an atom (Fig. 4.3): within the same shell the electrons occupy different orbitals with their spins aligned if possible, and then they start doubly occupying the orbitals.

The effect of Hund's coupling in magnetic insulators is also well known. What is becoming clear in recent years is that it also plays a crucial role in tuning the correlations in correlated metals (and their tendency to become Mott insulators) [30], and this physics, explored with the slave-spin mean-field is the subject of the following chapters.

In many practical applications only the three density-density terms of the Kanamori interaction are retained (dubbed interaction *à la* "Ising"). This introduces somewhat quantitative, but rarely qualitative changes to the results and that is what we will do for the rest of the manuscript.

4.6.1 Complement: *Particle-Hole Symmetry in the Multi-orbital Hubbard Model*

Applying the particle-hole symmetry to the Kanamori Hamiltonian for two orbitals one obtains the value of the chemical potential yielding half-filling in cases where the hopping structure is particle-hole symmetric, as done for the single-band Hubbard model in complement Sect. 4.3.1.

It can also be obtained by simply writing the Hamiltonian in a manifestly invariant form and then extract the chemical potential shift needed to obtain it from

$$\hat{H}_{int} - \mu \sum_{m\sigma} n_{m\sigma}, \tag{4.94}$$

where $\hat{H}_{int}$ is of the form (4.92). Indeed we can write:

$$\hat{\hat{H}}_{int} = U \sum_{m} \tilde{n}_{m\uparrow}\tilde{n}_{m\downarrow} + U' \sum_{m\neq m'} \tilde{n}_{m\uparrow}\tilde{n}_{m'\downarrow} + (U' - J) \sum_{m<m',\sigma} \tilde{n}_{m\sigma}\tilde{n}_{m'\sigma} \tag{4.95}$$

where all number operators have been shifted by 1/2 using the notation $\tilde{n}_{m\sigma} \equiv n_{m\sigma} - 1/2$, which means that we have only added one-body terms, and a constant energy shift. We have also dropped the non density-density terms of (4.92) (the so-called "spin-flip" and "pair-hopping" terms), which are manifestly particle-hole invariant and hence do not introduce any chemical potential shift under particle-hole transformation.

Developing the expression (4.95) and matching it with (4.94) (dropping the non-diagonal terms there too) one finds that $\mu_{1/2 filling} = \frac{U(2N-1)-5J(N-1)}{2}$.

4.6.2 *Extension of the Slave-Spin Formalism to Multi-orbital Hubbard Models*

In the multi-orbital context the Hubbard model becomes:

$$\hat{H} = \sum_{i\neq j mm'\sigma} t_{ij}^{mm'} d_{im\sigma}^{\dagger} d_{jm'\sigma} + \sum_{im\sigma} \epsilon_m n_{im\sigma}^{d} + \hat{H}_{int} \tag{4.96}$$

where $\hat{H}_{int}$ is of the form (4.92). We have also written explicitly the on-site one-body terms, that are the orbital energies $t_{ii}^{mm} \equiv \epsilon_m$, and we have made the hypothesis that there are no off-diagonal on-site hopping terms (local hybridizations), which is always the case in the Wannier function basis for a proper unit cell respecting the symmetries of the lattice.

The Slave-spin formalism is easily extended to this multi-orbital case. Indeed already in the one-band case every fermionic variable is supplemented with a corresponding auxiliary spin-1/2 variable, so that these spin variables were labeled, as the fermionic ones by site and spin indices.

Now the fermions have also an orbital index $d_{im\sigma}$, and so will the auxiliary spins $S^z_{im\sigma}$ (and obviously the corresponding fermions of the enlarged Fock space $f_{im\sigma}$). The physical states are always individuated by the same constraint (4.56), which is diagonal in all the indices so it is easily generalized:

$$\langle n^f_{im\sigma}\rangle_f = \langle S^z_{im\sigma}\rangle_s + \frac{1}{2}, \tag{4.97}$$

and is enforced at the mean-field level through orbital-dependent Lagrange multipliers λ_m.

The interaction has now further terms. As already mentioned we will most often drop the spin-flip and pair-hopping terms, which necessitate by the way an approximate treatment at the present stage of development of the slave-spin formalism [21].[9] We will then use the interaction in the "Ising" form (4.95), which can easily be expressed with the corresponding z-component of the slave-spins, as in the one-band case. That is (4.95) is expressed as:

$$\hat{\tilde{H}}_{int}[S] = U\sum_m S^z_{m\uparrow}S^z_{m\downarrow} + U'\sum_{m\neq m'} S^z_{m\uparrow}\tilde{S}^z_{m'\downarrow} + (U'-J)\sum_{m<m',\sigma} S^z_{m\sigma}S^z_{m'\sigma}. \tag{4.98}$$

The hopping term remains of the form (4.49) but now all terms carry orbital indices in which they can be off-diagonal. The mean-field decoupling is strictly analogous, so that the final mean-field equations are:

$$H_f = \sum_{i\neq j,mm'\sigma} t^{mm'}_{ij}\sqrt{Z_m Z_{m'}}\, f^\dagger_{im\sigma}f_{jm'\sigma} + \sum_{im\sigma}(\epsilon_m - \lambda_m - \mu)n^f_{im\sigma}, \tag{4.99}$$

$$H_s = \sum_{m,\sigma}\left[(h_m O^\dagger_{m\sigma} + H.c.) + \lambda_m(S^z_{m\sigma} + \frac{1}{2})\right] + \hat{\tilde{H}}_{int}[S], \tag{4.100}$$

$$h_{m\sigma} = \sum_{m'}\langle O_{m'\sigma}\rangle_s \sum_{j(\neq i)} t^{mm'}_{ij}\langle f^\dagger_{im\sigma}f_{jm'\sigma}\rangle_f \tag{4.101}$$

$$Z_m = |\langle O_{m\sigma}\rangle|^2 \tag{4.102}$$

together with (4.97).

The gauge number $c_{im\sigma}$ of (4.61) and treated for the single-band case in the complement Sect. 4.5.2, is now orbital dependent, keeping the same dependence on the

[9] Where they have been used [21, 31–33] the spin-flip and pair-hopping terms have been approximated using only slave-spin operators as $-J(S^+_{1\uparrow}S^-_{1\downarrow}S^+_{2\downarrow}S^-_{2\uparrow} + S^+_{1\uparrow}S^+_{1\downarrow}S^-_{2\uparrow}S^-_{2\downarrow} + h.c.)$, which, even if they do not flip the fermionic part as they should, have the virtue of reproducing the atomic spectrum of the Kanamori hamiltonian. The results are satisfactory in the two-band case, whereas major discrepancies are found for $N \geq 3$ for the non-half-non-singly filled cases (the "Janus" cases detailed in Sect. 4.6.7).

orbital density (provided, as already assumed here, the absence of local hybridizations), i.e.,:

$$c_{im\sigma} = c_m = \left(1/\sqrt{\langle n^f_{im\sigma}\rangle(1-\langle n^f_{im\sigma}\rangle)} - 1\right) \tag{4.103}$$

This choice, as again in the single-band case, leads to the correct values $Z_m = 1$ in the noninteracting limit, and it is thus what we adopt.

One slight problem in this formulation is that it yields nonzero Lagrange multipliers in the noninteracting limit, when a crystal-field splitting is present, which is an annoying feature. However it is not fundamentally problematic. We then compensate this by shifting the local energies ϵ_m to $\tilde{\epsilon}_m$ so that $\tilde{\epsilon}_m - \lambda_m(\{U, J\} = 0, \{\tilde{\epsilon}_m\}) = \epsilon_m$ in order to obtain the correct non-interacting population distribution among the orbitals at zero interaction strength.

Since we have used the particle-hole symmetric form of the interaction, here the chemical potential that leads to half-filling for a particle-hole symmetric DOS is $\mu = 0$.

4.6.3 *The N-Orbital Hubbard Model*

For the particular choice $J = 0$ and for identical diagonal (in the orbital index) hopping integrals, the multi-band Hubbard model acquires the form:

$$\hat{H} = \sum_{ijm\sigma} t^m_{ij} d^\dagger_{im\sigma} d_{jm\sigma} + \frac{U}{2}\sum_i \left(\sum_{m\sigma}\left(n^d_{im\sigma} - \frac{1}{2}\right)\right)^2, \tag{4.104}$$

which is $SU(2N)$ symmetrical for rotations in the local spin-orbital space. Indeed the interaction term is nothing else than $U\sum_i (Q_i - N/2)^2$, where Q_i is the total charge on site i, which is independent of the local basis chosen.

The Mott transition can intuitively happen for any integer number of electrons per atomic site, since what is needed is the blocking of the motion of the electrons due to their mutual repulsion, which can only work when there are no "spare" sites to move to.

Indeed this intuitive result is found in multi-orbital Hubbard models through several techniques. In Fig. 4.4 the result from [34] in which the Gutzwiller approximation is used. This approximation coincides with the Kotliar–Ruckenstein slave-bosons [20], which in many instances gives identical results with the slave-spin mean field. The results in Fig. 4.4 can be considered representative of SSMF.

The critical interaction strength for the Mott transition is in general a growing function of the atomic ground state degeneracy (as shown for example in complements Sects. 4.5.5 and 4.6.4), which implies, in presence of the simple atomic hamiltonian (4.104) with $J = 0$, a larger U_c for a higher number of orbitals, and for fillings closer to half, as is clear from the figure.

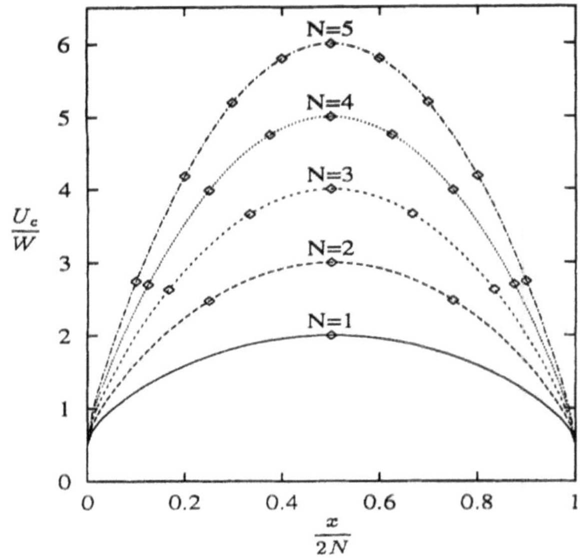

Fig. 4.4 Critical interaction strengths U_c for the Mott transition in the SU(2N) symmetrical (i.e., with zero Hund's coupling) Hubbard model of bandwidth W, as a function of the filling (here x denotes the density per spin). Mott transitions (*diamonds*) happen for all the integer fillings allowed, and the U_c grows in general with the number of orbitals and the proximity to half-filling, in this $J = 0$ case. From [34]

Indeed these results are confirmed within other approximation schemes, like DMFT [35, 36].

The exact large-N behaviour expected for the critical interaction U_c at half-filling was calculated in [37] and reads:

$$\frac{U_c}{N} = 8|\bar{\epsilon}|, \qquad \text{(at large N)} \tag{4.105}$$

The same perturbative technique used in the complement Sect. 4.5.5 can be used to determine the result from the Slave-spin technique (see complement Sect. 4.6.4), that reads:

$$U_c = 8(N+1)|\bar{\epsilon}| \tag{4.106}$$

and is obviously confirmed by the numerical results reported in Fig. 4.5.

This shows then that the Slave-spin mean-field captures exactly the large-N critical interaction strength for the multi-orbital Hubbard model.

In Fig. 4.5 the numerical calculation within SSMF of the quasiparticle weight as a function of the interaction strength U/D in this model is compared for different number of orbitals $N = 1, 2, 3, 4$.

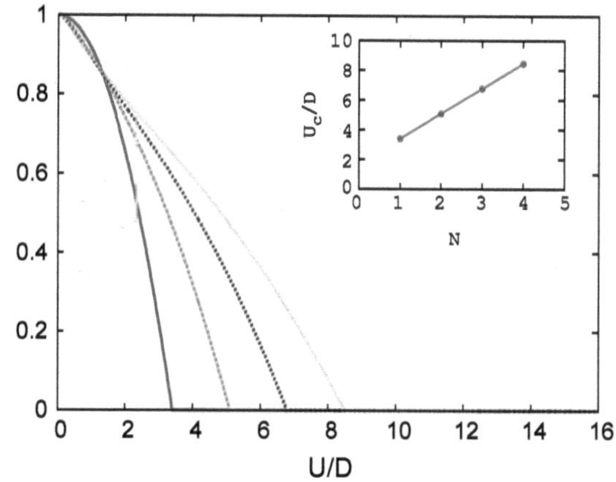

Fig. 4.5 Quasiparticle weight, obtained from slave-spin mean-field for the N-orbital Hubbard model at half-filling (with, from *left to right* $N = 1, 2, 3, 4$). The non-interacting density of states is a semi-circle with half bandwidth D *Inset* Dependence of the critical U on N. The exact large- N behaviour [37] is obtained (see text). From [21]

4.6.4 Complement: *Critical U for the Mott Transition in the 2-Orbital Hubbard Model*

Here we show how to obtain analytically within the slave-spin mean-field approximation the critical interaction strength U_c for the Mott transition in the half-filled (particle-hole symmetric) 2-band Hubbard model in absence of Hund's coupling ($J = 0$). The same technique can yield analogously the U_c for any number N of orbitals, and reproduce exactly the numerical results shown in Fig. 4.5. The case for finite J is developed in complement Sect. 4.6.9.

As done in the complement Sect. 4.5.5, the multi-orbital slave-spin equation (4.100) can be developed around the insulating solution $h_{im\sigma} = 0$ where $H_s = H_{at}$ with, for this multi-orbital case with $J = 0$:

$$H_{at} = \frac{U}{2}\left(\sum_{m\sigma} S^z_{m\sigma}\right)^2, \tag{4.107}$$

of which we know the spectrum. This is:

$$\begin{cases} |\uparrow\uparrow,\uparrow\uparrow\rangle |\downarrow\downarrow,\downarrow\downarrow\rangle & E = 2U \\ \\ |\uparrow\uparrow,\uparrow\downarrow\rangle |\uparrow\uparrow,\downarrow\uparrow\rangle |\downarrow\downarrow,\uparrow\downarrow\rangle |\downarrow\downarrow,\downarrow\uparrow\rangle & \\ |\uparrow\downarrow,\uparrow\uparrow\rangle |\downarrow\uparrow,\uparrow\uparrow\rangle |\uparrow\downarrow,\downarrow\downarrow\rangle |\downarrow\uparrow,\downarrow\downarrow\rangle & E = \frac{U}{2} \\ \\ |\uparrow\downarrow,\uparrow\downarrow\rangle |\uparrow\downarrow,\downarrow\uparrow\rangle |\downarrow\uparrow,\uparrow\downarrow\rangle |\downarrow\uparrow,\downarrow\uparrow\rangle & \\ |\uparrow\uparrow,\downarrow\downarrow\rangle |\downarrow\downarrow,\uparrow\uparrow\rangle & E = 0 \end{cases} \tag{4.108}$$

We again calculate as a perturbation, the effect of the "kinetic" part of the slave-spin hamiltonian:

$$H_{pert} = h \sum_{m\sigma} 2S^x_{m\sigma}. \tag{4.109}$$

where $O_{im\sigma} = 2S^x_{m\sigma}$ in (4.100) because both orbitals are half-filled and thus the gauge $c = 1$ ((4.61) for $n_m = 1/2$), and we call $h \equiv h^*_{im\sigma} + h_{im\sigma}$. Also $\lambda_m = 0$ because of particle-hole symmetry.

As for the single-band case we perform a perturbative expansion in h, which is small near the metal-insulator transition, if this is second-order.

In this $J = 0$ case the ground state has degeneracy 6 (the subset above with $E = 0$). Furthermore, as before, the perturbation H_{pert} does not have any nonzero elements in the low-energy subspace one has to use second-order perturbation theory to find the right combination of states, in the ground state manifold, to which the perturbed state tends for $h \to 0$. This is the lowest-lying eigenstate of the matrix $H' \equiv H_{pert}(E_0 - H_{at})^{-1} H_{pert}$ in the degenerate subspace. In this case, since H_{pert} couples the $E = 0$ sector only with the $E = \frac{U}{2}$ sector, one can rewrite $H' = -\frac{2}{U} H^2_{pert}$ (where we have used the fact that $E_0 - H_{at}$ is diagonal and $E_0 = 0$).

The complete 6×6 matrix for the restriction of H_{pert} to the $E = 0$ subspace to be diagonalized is reported in Appendix ((4.136) with $h_1 = h_2$). However by inspection one can easily convince himself that

$$|\phi_0\rangle \equiv \frac{1}{\sqrt{6}}(|\uparrow\downarrow,\uparrow\downarrow\rangle + |\uparrow\downarrow,\downarrow\uparrow\rangle + |\downarrow\uparrow,\uparrow\downarrow\rangle + |\downarrow\uparrow,\downarrow\uparrow\rangle + |\uparrow\uparrow,\downarrow\downarrow\rangle + |\downarrow\downarrow,\uparrow\uparrow\rangle) \tag{4.110}$$

is the ground state to leading order in h. Indeed H^2_{pert} has the effect of flipping any two spins (including twice the same), so that any state of the $E = 0$ manifold is changed, when acted upon by V^2, into itself plus four of the others states of the manifold (only one is left out, since it needs 4 flips to be reached). Then it is clear that the ground state is the symmetric sum of all the 6 states since it is obviously changed into itself by H^2_{pert} and the plus signs in the sum add up all the $-\frac{2}{U}$ contributions, obtaining the lowest possible eigenvalue.

Once found the correct zeroth-order ground state $|\phi_0\rangle$ we need to find its ket correction to leading order in h.

The correction to the ket is nonzero at first order in h and reads:

$$|\phi_0^{(I)}\rangle = |\phi_0\rangle + \sum_{|s\rangle \neq |\phi_0\rangle} \frac{\langle s|H_{pert}|\phi_0\rangle}{E^0 - E^0_s} |s\rangle = |\phi_0\rangle - \frac{6h}{\sqrt{6}U} \sum_{|s\rangle \subset \{E=\frac{U}{2}\}} |s\rangle, \tag{4.111}$$

because $\langle s|H_{pert}|\phi_0\rangle = \frac{3h}{\sqrt{6}}$ for any $|s\rangle \subset \{E = \frac{U}{2}\}$.

Analogously to the 1-band case to calculate the critical interaction one uses the self-consistency condition $h = 2\bar{\epsilon}\langle 2S^x_{m\sigma}\rangle$. This condition has to be rigorously satisfied and since we can only calculate here a perturbative development for $\langle 2S^x_{m\sigma}\rangle$, which by definition is only exactly valid at the transition, (where $h \to 0$), the self-consistency condition becomes an equation determining the coupling U for which this condition can be satisfied rigorously, hence the critical coupling U_c.

Applying $2S^x_{m\sigma}$ to the ground state:

$$2S^x_{1\uparrow}|\phi_0^{(I)}\rangle = 2S^x_{1\uparrow}\left(\frac{1}{\sqrt{6}}\sum_{|s\rangle\subset\{E=0\}}|s\rangle - \frac{\sqrt{6}h}{U}\sum_{|s'\rangle\subset\{E=\frac{U}{2}\}}|s'\rangle\right) \tag{4.112}$$

one obtains 6 of the 8 states of the $E = \frac{U}{2}$ manifold from the acting of $2S^x_{1\uparrow}$ on the $|s\rangle$ states and the 6 states of the $E = 0$ manifold plus the two from the $E = 2U$ one, from the acting on of $2S^x_{1\uparrow}$ on the $|s'\rangle$ states, which implies that

$$\langle\phi_0^{(I)}|2S^x_{1\uparrow}|\phi_0^{(I)}\rangle = -\frac{12h}{U}. \tag{4.113}$$

The self-consistency equation becomes then:

$$h = 2\bar{\epsilon}\langle 2S^x_{m\sigma}\rangle = -\frac{24h}{U_c}\bar{\epsilon} \implies U_c = -24\bar{\epsilon}. \tag{4.114}$$

For a semi-circular DOS of half-bandwidth D ($\mu = \lambda_m = 0$ because of particle-hole symmetry), $\bar{\epsilon} \simeq -0.2122D$, which gives

$$U_c \simeq 5.09 \tag{4.115}$$

if $D = 1$, in agreement with the numerical result Fig. 4.5.

4.6.5 Effect of Hund's Coupling on the Mott Transition

We will now see what is the effect of a finite Hund's coupling $J \neq 0$ on the critical interaction strength needed to obtain a Mott insulator, and in general the influence that this has on the degree of correlation of the neighboring metallic phases.

Figure 4.6 shows the quasiparticle weight as a function of the interaction strength, for various fixed values of J/U, in the 2-orbital Hubbard model at half-filling.

The two forms Hund's coupling are explored here: Kanamori (approximately treated, as indicated in footnote 9 on p. xxx) and Ising. In both cases Hund's coupling is seen to reduce the U_c for the Mott transition. Besides a stronger effect from the Ising form (and consequently a more correlated metal at the same value of J/U compared to the Kanamori form), the main difference that is noticed is on the order of the transition. Indeed the transition is second order at J = 0. The onset of J in any form makes it immediately first order. However in the Kanamori form the second order of the transition is restored above $J/U \simeq 0.18$ (as signaled by the vanishing jump in Z at the transition). For the Ising form instead the first order character persists until the largest physical value $J/U = 1/3$ and is always stronger for larger J/U. Despite the approximate treatment of the Kanamori Hamiltonian this non-monotonous behaviour of the first-order jump is qualitatively well captured by SSMF, as can be shown by

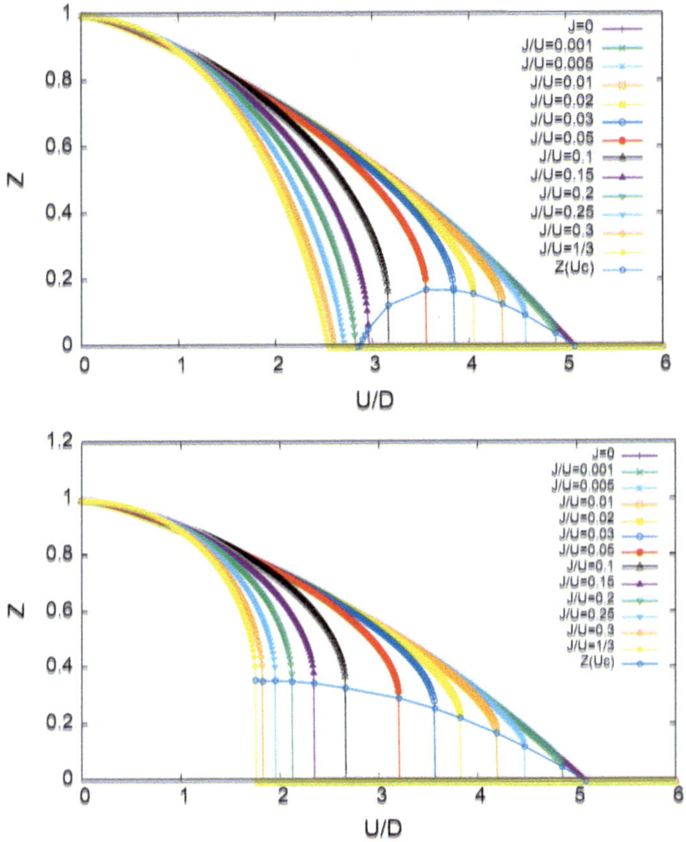

Fig. 4.6 Quasiparticle weight (also the inverse mass enhancement, in local schemes like SSMF) calculated for the half-filled 2-orbital Hubbard model as a function of the interaction strength U/D for various fixed values J/U. In this half-filled case one sees that the critical interaction strength for the Mott transition is reduced by an increasing Hund's coupling. *Upper panel* Kanamori form of Hund's coupling. *Lower panel* Ising form

comparing to other techniques like rotationally-invariant slave bosons (RISB) [38] or DMFT [39].

An opposite effect of Hund's coupling is found instead in the quarter-filled 2-orbital Hubbard model. Indeed as illustrated in Fig. 4.7, a finite Hund's coupling decorrelates the system in this case, and pushes the U_c for the Mott transition at very high interaction strengths.

The main cause of this opposite behaviour in the half-filled and quarter-filled cases in the 2-band model is understood from the energetics of the atomic Mott gap.

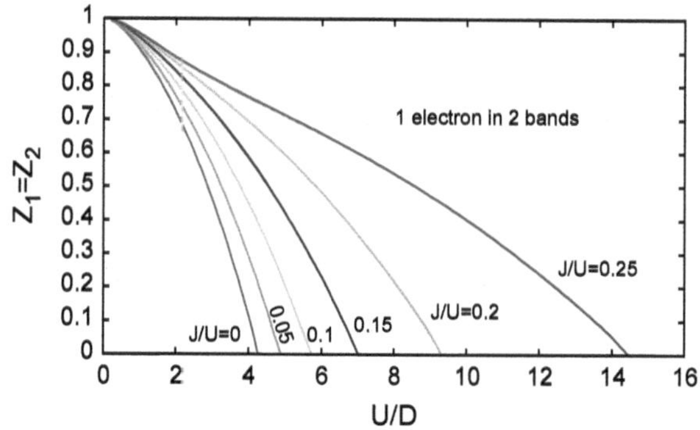

Fig. 4.7 Quasiparticle weight calculated in SSMF for a quarter-filled 2-orbital Hubbard model as a function of the interaction strength for various Hund's coupling strengths. From [33]

4.6.6 The Physics of the Mott Gap

Indeed the excitation spectrum of an idealized paramagnetic Mott insulator (such as the one obtained at large U in the Hubbard hamiltonian (4.104)) is easily put in correspondence with the atomic one. The so-called upper and lower Hubbard bands are the two main spectral features due to the addition and removal of a particle respectively, and correspond to the atomic transitions that one would obtain in the "atomic limit", i.e., putting all hoppings to zero in (4.104). In such an isolated Hubbard atom the eigenstates are the local states with definite charge n and the distance between the ground state and the excited states with charge $n \pm 1$ is $U/2$, in absence of Hund's coupling. In the lattice model these excited states are as many as the lattice sites (only one site has a ± 1 charge) times the local degeneracy due to orbital and spin indices, and are obviously degenerate. Reintroducing the hopping (for example as a perturbation) lifts this degeneracy and disperses the eigenstates on an energy range of the order of the bandwidth, thus forming the Hubbard bands.

From this point of view an intuitive criterion, due to J. Hubbard himself [40], for the breakdown of the Mott insulating state can easily be obtained. Indeed the distance between the Hubbard bands is set by U (in the particle-hole symmetric case they are centered at energy $\pm U/2$) while their dispersion is set by the bandwidth W (and is $\sim eq W$ in the single-band case, while it can be different, but always of order W, in multi-band cases, as it will be shown in the next section), so that at large U/W an energy gap exists in both the addition and subtraction spectrum. This reflects the energy cost of moving a particle in the Mott insulator: it amounts to the energy of creating, on the ground state with all sites with n particles, a site with an extra particle, thus leaving behind a site with an extra hole, minus the energy that can be gained with the delocalization of these particles. When this cost is zero, the Mott insulator cannot exist, and a metal is obtained (in absence of other symmetry breakings) which by definition is the state with delocalized excitations at vanishing energy.

Quantitatively this means that the Mott gap is the atomic excitation gap minus the bandwidth, i.e., (in the single-band case) $\Delta_{Mott} = \Delta_{at} - W$, and since $\Delta_{at} \equiv E_{at}(n+1) + E_{at}(n-1) - 2E_{at}(n) = U$ in absence of Hund's coupling the Hubbard criterion for its closure reads:

$$\Delta_{Mott} = U - W = 0 \Longrightarrow U_c = W \tag{4.116}$$

This reflect the natural physical expectation that a Mott insulator is obtained when the Coulomb repulsion between the electrons overcomes the delocalization energy due to the hopping.

Now, the atomic excitation spectrum is modified by a finite Hund's coupling J and so is the atomic Mott gap, thus implying that the Hubbard criterion yields $U_c = U_c(J)$. Indeed when H_{int} takes the form (4.92) it is easily shown [33] that the atomic Mott gap becomes

$$\Delta_n^{at} = \begin{cases} U - 3J, & \forall n \neq N \quad \text{(off half-filling)}, \\ U + (N-1)J, & n = N \quad \text{(half-filling)}. \end{cases} \tag{4.117}$$

The atomic Mott gap is thus increased by J in the half-filled case, and reduced in all other cases. This implies that J favors the Mott insulator at half-filling, thus reducing the corresponding U_c, and disfavors it for all other fillings, moving thus the U_c to very high values. Indeed in Fig. 4.8 the numerical results from SSMF are reported for all integer fillings in both the 2-orbital and the 3-orbital Hubbard models. It is clear that at large J these results are well captured by the simple Hubbard criterion with Δ^{at} from formulas (4.117).

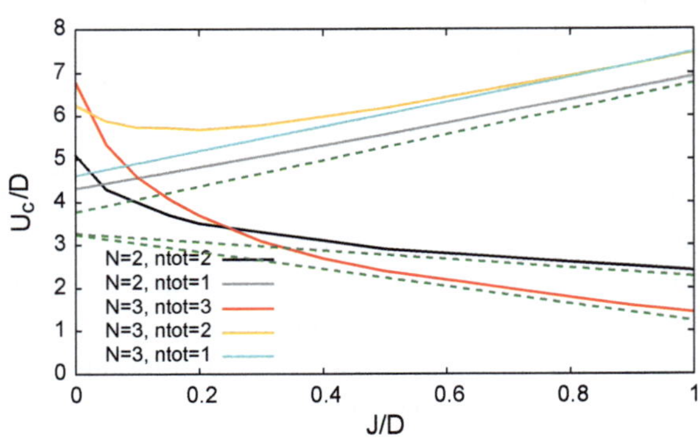

Fig. 4.8 Critical interaction strength for the Mott transition for all integer fillings (fillings for $n > N$ yield identical results because of the particle-hole symmetry) in the 2-orbital and 3-orbital Hubbard model, as a function of Hund's coupling J. The *dashed lines* indicate the large-J asymptotics based on the dependence on J of the atomic gap, formulas (4.117) (from *top to bottom* $\delta U_c \propto 3J, -J, -2J$). From [33]

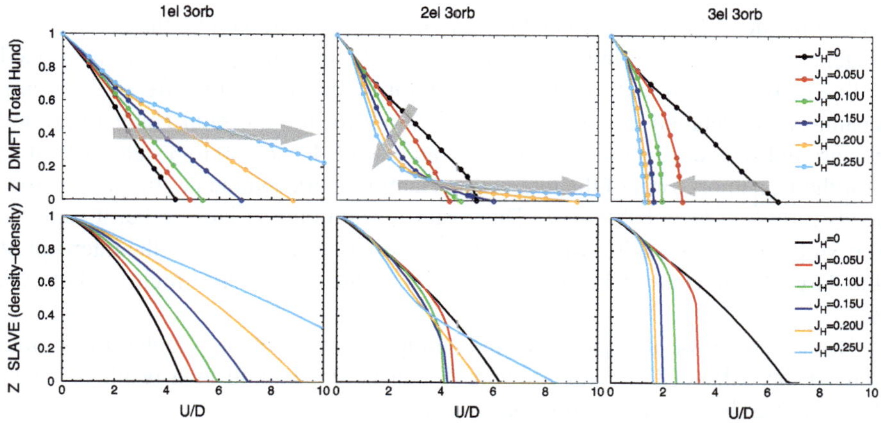

Fig. 4.9 Comparison of the quasiparticle weight and inverse mass enhancement Z calculated in DMFT (Kanamori Hamiltonian, *top panels*) and SSMF (Ising hamiltonian, *bottom panels*) in the 3-band Hubbard model, for various values of J/U. The *grey arrows* indicate schematically the effect of a growing Hund's coupling J/U. Adapted from [42] (DMFT data from [41])

4.6.7 Hund's Effect on Orbital Fluctuations and the Narrowing of the Hubbard Bands

The data plotted in Fig. 4.8 show that at small J another effect of Hund's coupling is visible, that is a fast reduction of U_c from the high values reached at $J = 0$ in all cases but those in which the system is filled by one electron or one hole per site only. This reduction clearly adds up with the large-J behaviour, thus causing a superlinear reduction of U_c in the half-filled cases and a non-monotonous $U_c(J)$ in the generic non-half-non-singly-filled cases (which are present in models from N = 3 orbitals up).

In Fig. 4.9 the calculation (performed in DMFT [41] and SSMF [42]) of the quasiparticle weight in the 3-orbital Hubbard model is reported, showing in detail the aforementioned two effects clearly at play. Indeed the central panel shows the non-half-non-singly-filled case which was dubbed "Janus-faced", because there the two effects of Hund's coupling play an opposite role[10]: the shifting of U_c towards high values prevails at large J, but an overall reduction of the quasiparticle weight and thus an increase in the correlation of the system with J happens for a large range of U and even causes the U_c to decrease initially.

Intuitively this enhancement of correlation at small J can be viewed as the opposite of the effect reducing the correlation with the number of orbitals N at $J = 0$, and ultimately leading to an increase in $U_c^N(J = 0)$ with N. This effect was treated in Sect. 4.6.3 and stems from the systematic increase of the ground-state degeneracy and of the hopping channels with N, as also illustrated explicitly by the comparison

[10] Janus, the ancient roman god of beginnings, transitions and endings and of time, gates, doors etc., was depicted with two faces looking in opposite directions.

of complements Sects. 4.5.5 and 4.6.4. A finite J, however, reduces the ground-state degeneracy again, lowering it to that of the high-spin multiplet (for a given fixed filling of n electrons in N orbitals). This in turn reduces also the effective hopping channels. As detailed in complement Sect. 4.6.9 for the 2-orbital Hubbard model, the ground state degeneracy is brought from 6 states for $J = 0$ to 3 states at finite J (for the Kanamori form of the interaction Hamiltonian), and the perturbation theory leads to $U_c^{N=2}(J) = U_c^{N=1} - J$. This result is valid only at large J, but it captures the Mott gap linear scaling with J, and interestingly also shows that the intercept of this large-J behaviour at J = 0 is the critical interaction for the single-orbital case $U_c^{N=1}$.

This fact can be interpreted in the simplified terms of the Hubbard criterion. Indeed the enhanced U_c^N for growing N can be understood as a result of larger Hubbard bands, so that the criterion becomes:

$$\Delta_{Mott} = \Delta_{at} - \tilde{W} \tag{4.118}$$

where $\tilde{W}$ is the width of the Hubbard bands, that was indeed found to increase with the number of orbitals in [43, 44] as $\tilde{W} \simeq W\sqrt{N}$, due to the orbital quantum fluctuations in the ground state. The criterion for the closure of the gap $\Delta_{Mott} = 0$ then implies ($\Delta_{at} = U$ also in the multi-orbital case) that $U_c^N = \tilde{W} \simeq \sqrt{N} U_c^{N=1}$.

This result is qualitatively in agreement with the increase of $U_c^N \propto N$ treated in Sect. 4.6.3. The quantitative disagreement (N vs. $\sqrt{N}$) is explained by an increasingly larger coexistence zone between an insulator and a metal with a preformed gap. Indeed two spinodal lines (where the insulating and the metallic solutions disappear, respectively) are found corresponding to the Mott transition in DMFT. They give rise to a zone of coexistence of solutions, in which the actual transition is located exactly by comparing the energy of the two solutions [45]. This zone is seen to widen with N at $J = 0$ [37]. However there are indications that this coexistence zone strongly narrows when J is nonzero [46]. We will thus ignore to distinguish precisely between the two lines, even if conceptually we are making a leap, since the SSMF traces the disappearance of the metallic solution while the Hubbard criterion describes the closure of the insulating gap. Quantitatively however, at finite J, this discrepancy is a small one.

Thus by making use, for the finite-J case, of the second of (4.117) (relevant for this half-filled case at $N = 2$) $\Delta^{at} = U + J$, in (4.118) one finds, for the closure condition of the Mott gap, $U_c = \tilde{W} - J$. This compared with large-J behaviour $U_c^{N=2}(J) = U_c^{N=1} - J$ derived for the 2-orbital Hubbard model in complement Sect. 4.6.9, signals then that the width of the Hubbard bands is reduced by Hund's coupling back to a value of the order of the one-band case $\tilde{W} = W$.[11]

[11]The precise result $\tilde{W} = W$ should be taken with some care however, since our treatment of the Kanamori hamiltonian is approximate. Indeed when deriving the same perturbative result with the Ising form of Hund's coupling one finds $U_c^{N=2}(J) = U_c^{N=1}/2 - J$. The Ising hamiltonian is treated exactly, however the Mott transition is always first order, which invalidates the perturbative expansion. We take nevertheless these results together as indicating the large-J behaviour for U_c^N

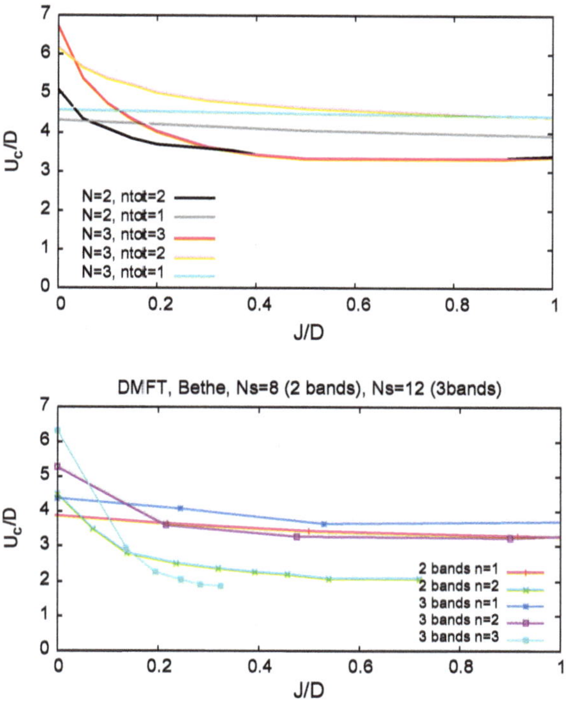

Fig. 4.10 Critical interactions for the Mott transition in the 2-band and 3-band Hubbard models compensated for the J-dependence of the Mott gap. Specifically J and 2 J have been added to the N = 2 and N = 3 half-filled cases, respectively, while 3 J has been subtracted for all the other cases, following (4.117). *Upper panel* SSMF. *Lower panel* DMFT. In the half-filled cases this rescaled U_c, indicative of the width of the Hubbard bands (see text), is of the order of the bandwidth at finite J, while it is strongly enhanced for vanishing J

This result is confirmed in DMFT. Indeed Fig. 4.10 shows the U_c for all the Mott transitions in the 2-band and 3-band degenerate Hubbard models as a function of J, and compares the result obtained by SSMF and DMFT, with the Kanamori Hamiltonian. In the plotted $U_c(J)$ a linear contribution in J is added or subtracted, following (4.117), to compensate the effect due to the J-dependence of the Mott gap, and thus highlighting the remaining effect of reduction by J of the orbital fluctuations in the ground state, and of the width of the Hubbard bands.

Indeed the very argument used in [43, 44] to support the enhanced width of the Hubbard bands at J = 0, was shown to break down by Koga et al. in [47]. The argument is essentially that Hund's coupling suppresses the orbital fluctuations responsible for the enhancement.

The width of the Hubbard bands can be also assessed in a direct manner within DMFT, where an accurate local spectral function can be calculated. This result is reported in Fig. 4.11 for a 2-orbital Hubbard model, and confirms the conclusion that we have anticipated. Indeed the width of the Hubbard bands diminishes upon the

(Footnote 11 continued)
having a $J = 0$ intercept of the order of the single-band U_c. See also the discussion at the end of complement Sect. 4.6.9.

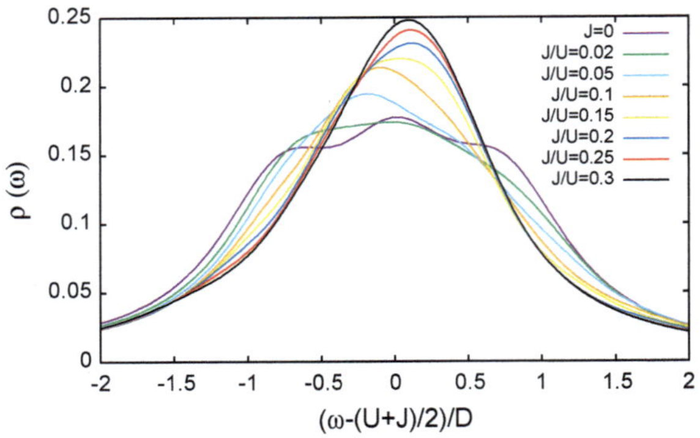

Fig. 4.11 Hubbard bands in the local spectral function $\rho(\omega)$ for the half-filled 2-orbital Hubbard model as a function of J (Kanamori Hamiltonian) in DMFT. The interaction is $U = 7D$, so that at all J/U the system is in a Mott insulating phase, and a gap is open in the spectrum. As for Fig. 4.10, following (4.117), a scaling of $(U + J)/2$ is performed (here on the abscissas), that brings the position of the atomic-limit excitation (roughly the barycenter of the Hubbard bands in the lattice model) to coincide for all J/U. Here we show thus the Hubbard band of the addition spectrum (positive frequencies). The result shows directly that the Hubbard bands have a width $\sim W = 2D$ at large J/U, reduced from the $J = 0$ value $\sim\sqrt{2}W$

onset of J, and it goes roughly from a value of order $\sim\sqrt{2}W$ at $J = 0$ to a value roughly of order $\sim W$ at large J/U.[12]

In agreement with this result another complementary analysis can be done in this framework. Indeed starting from a half-filled Mott insulator at large U, the chemical potential can be shifted in the gap until the edge of the gap itself. When the chemical potential is shifted further one recovers a metal with a finite density of states at the Fermi level. By moving then the chemical potential in the opposite direction there is another value of the chemical potential where this density of states vanishes and the chemical potential is within the gap again. This cycle defines the border of the Hubbard band within an error bar mainly due to the coexistence zone found in DMFT. The result of this procedure as a function of J (not shown, due to large error bars caused by the extremely low coherence scale of the metal in this regime that makes the convergence of the algorithm very hard at zero temperature) confirms that the Hubbard bands are roughly narrowed down by J from a value $\sim\sqrt{N}W$ to one of $\sim W$, and that the coexistence zone shrinks sizably.

[12]The uncertainty on the numerical result is due to the fact that exact diagonalization of a finite bath in the auxiliary impurity system used in the DMFT scheme is used for the calculation of the local spectral function. The discretization implied by this impurity solver ($N_s = 10$ total sites between bath sites and the two-orbital impurity are used) yields a spiky spectral function that needs to be evaluated slightly away from the real axis in order to appear continuous. This broadening preserves the spectral weight distribution however, letting us conclude reliably on the width of the Hubbard bands.

4.6.8 Complement: *Comparison with DMFT*

In order to assess the accuracy of the SSMF a more detailed comparison is in order with results from the state-of-the-art technique for the analysis of local correlations in such Hubbard-like models, Dynamical Mean-Field Theory.

The upper panel of Fig. 4.12 shows such a comparison for the U_c for the Mott transition in the half-filled 2-orbital Hubbard model with both Ising and Kanamori interaction hamiltonians. It can be notices that SSMF slightly but systematically overestimates the U_c compared to DMFT. DMFT also confirms the stronger correlations induced by the Ising interaction, as highlighted in Fig. 4.6.

It is to be noticed that the SSMF treatment of the Ising hamiltonian (of which the mapping in the auxiliary Hilbert space is exact) gives results reasonably close those for the Kanamori hamiltonian treated in DMFT. Indeed in both the 2-band and the 3-band half-filled models results for typical physical values of $J/U \simeq 0.12 \div 0.15$ for the DMFT treatment of the Kanamori hamiltonian are reproduced by the SSMF treatment of the Ising hamiltonian with $J/U \simeq 0.2 \div 0.25$.

This estimate is confirmed in more detail in the comparison performed in [42] and reported in Fig. 4.9.

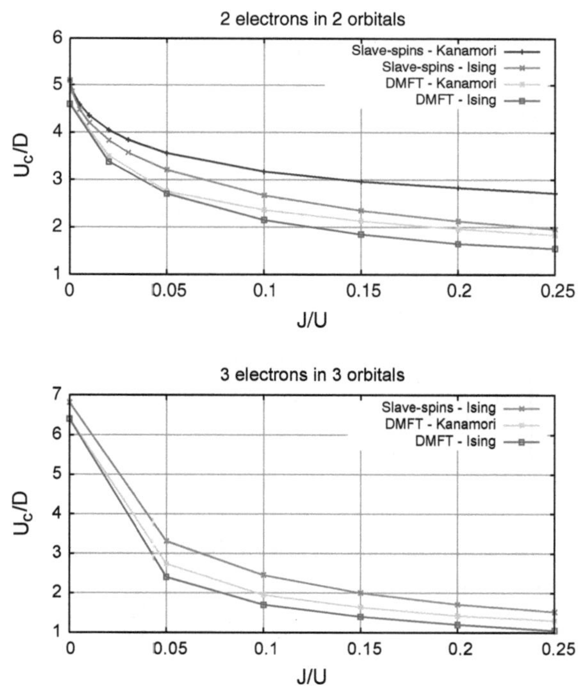

Fig. 4.12 Comparison of the SSMF critical interaction strength for the Mott transition in the half-filled 2-band (*upper panel*) and 3-band (*lower panel*) Hubbard models with both Ising and Kanamori interactions with the analogous results in DMFT. A similar comparison for the Mott transition in the quarter-filled 2-band Hubbard model with Kanamori interaction is reported in [33] (Fig. 6, lower panel)

4.6.9 Complement: *Critical U for the Mott Transition in the 2-Orbital Hubbard Model for Large Hund's Coupling J, and Orbital Decoupling*

Here we show how to obtain analytically within the slave-spin mean-field approximation the critical interaction strength U_c for the Mott transition in the (particle-hole symmetric) half-filled 2-band Hubbard model in presence of Hund's coupling ($J \neq 0$). As shown by the numerical results (Fig. 4.6) the Mott transition in presence of Hund's coupling (in the Kanamori form) is second order only at large J. Thus the present results bear validity only at large J (which is why they do not connect with the analytical results obtained for $J = 0$ in Sect. 4.6.4.

We also show explicitly that in this large J limit the equations determining U_c for each band decouple, so that when the bandwidth of the two bands differs the U_c for the two band differs too, and a zone where an orbital-selective Mott phase is realized opens.

As done previously we develop the multi-orbital slave-spin equation (4.100) around the insulating solution $h_{im\sigma} = 0$ where $H_s = H_{at}$ with, now that $J \neq 0$:

$$\hat{\tilde{H}}_{int}[S] = U \sum_{m} S^z_{m\uparrow} S^z_{m\downarrow} + U' \sum_{m \neq m'} S^z_{m\uparrow} \tilde{S}^z_{m'\downarrow} + (U' - J) \sum_{m<m',\sigma} S^z_{m\sigma} S^z_{m'\sigma}$$
$$- J(S^+_{1\uparrow} S^-_{1\downarrow} S^+_{2\downarrow} S^-_{2\uparrow} + S^+_{1\uparrow} S^+_{1\downarrow} S^-_{2\uparrow} S^-_{2\downarrow} + h.c.) \quad (4.119)$$

of which we know the spectrum. This is[13]:

$$\begin{cases}
|\uparrow\uparrow,\uparrow\uparrow\rangle |\downarrow\downarrow,\downarrow\downarrow\rangle & E = 2U - 2J \\
|\uparrow\uparrow,\uparrow\downarrow\rangle \; |\uparrow\uparrow,\downarrow\uparrow\rangle \; |\downarrow\downarrow,\uparrow\downarrow\rangle \; |\downarrow\downarrow,\downarrow\uparrow\rangle & \\
|\uparrow\downarrow,\uparrow\uparrow\rangle \; |\downarrow\uparrow,\uparrow\uparrow\rangle \; |\uparrow\downarrow,\downarrow\downarrow\rangle \; |\downarrow\uparrow,\downarrow\downarrow\rangle & E = \frac{U+J}{2} \\
\frac{1}{\sqrt{2}}(|\uparrow\uparrow,\downarrow\downarrow\rangle - |\downarrow\downarrow,\uparrow\uparrow\rangle) & E = 4J \\
\frac{1}{\sqrt{2}}(|\uparrow\downarrow,\downarrow\uparrow\rangle + |\downarrow\uparrow,\uparrow\downarrow\rangle) \quad \frac{1}{\sqrt{2}}(|\uparrow\uparrow,\downarrow\downarrow\rangle + |\downarrow\downarrow,\uparrow\uparrow\rangle) & E = 2J \\
|\uparrow\downarrow,\uparrow\downarrow\rangle \quad \frac{1}{\sqrt{2}}(|\uparrow\downarrow,\downarrow\uparrow\rangle + |\downarrow\uparrow,\uparrow\downarrow\rangle) \quad |\downarrow\uparrow,\downarrow\uparrow\rangle & E = 0
\end{cases} \quad (4.120)$$

We again calculate as a perturbation, the effect of the "kinetic" part of the slave-spin hamiltonian:

$$H_{pert} = \sum_{m\sigma} h_m 2S^x_{m\sigma}. \quad (4.121)$$

[13] A constant energy shift (U+J)/2 was added everywhere to position the ground state on the zero of energy.

where $O_{im\sigma} = 2S^x_{m\sigma}$ because both orbitals are half-filled and thus the gauge $c = 1$ ((4.61) for $n_m = 1/2$), and we call $h_m \equiv h^*_{im\sigma} + h_{im\sigma}$. Also $\lambda_m = 0$ at half-filling because of particle-hole symmetry.

As done previously we perform a perturbative expansion in h_m, which is small near the metal-insulator transition, if this is second-order.

In this $J \neq 0$ case the ground state has degeneracy 3 (the subset above with $E = 0$). Furthermore, as before, the perturbation H_{pert} does not have any nonzero elements in the low-energy subspace one has to use second-order perturbation theory to find the right combination of states, in the ground state manifold, to which the perturbed state tends for $h_m \to 0$. This is the ground state of the matrix $H' \equiv H_{pert}(E_0 - H_{at})^{-1} H_{pert}$ in the degenerate subspace. In this case, since H_{pert} couples the $E = 0$ sector only with the $E = \frac{U+J}{2}$ sector, one can rewrite $H' = -\frac{2}{U+J} H^2_{pert}$ (where we have used the fact that $E_0 - H_{at}$ is diagonal and $E_0 = 0$).

By applying explicitly H' to the ground state triplet of states one finds the following 3×3 matrix (we keep here the h_m distinct):

$$H' = -\frac{4(h_1^2 + h_2^2)}{U + J} \begin{bmatrix} 1 & \frac{1}{\sqrt{2}} & 0 \\ \frac{1}{\sqrt{2}} & 1 & \frac{1}{\sqrt{2}} \\ 0 & \frac{1}{\sqrt{2}} & 1 \end{bmatrix} \tag{4.122}$$

of which the ground state to zero order in h_m is:

$$|\phi_0\rangle \equiv \frac{1}{2}(|\uparrow\downarrow, \uparrow\downarrow\rangle + |\uparrow\downarrow, \downarrow\uparrow\rangle + |\downarrow\uparrow, \uparrow\downarrow\rangle + |\downarrow\uparrow, \downarrow\uparrow\rangle). \tag{4.123}$$

The correction to the ground state to first order in h_m reads:

$$\begin{aligned} |\phi_0^{(I)}\rangle &= |\phi_0\rangle + \sum_{|s\rangle \neq |\phi_0\rangle} \frac{\langle s|H_{pert}|\phi_0\rangle}{E^0 - E^0_s} |s\rangle \\ &= |\phi_0\rangle - \frac{2}{U+J} [h_1(|\uparrow\uparrow, \uparrow\downarrow\rangle + |\uparrow\uparrow, \downarrow\uparrow\rangle + |\downarrow\downarrow, \uparrow\downarrow\rangle + |\downarrow\downarrow, \downarrow\uparrow\rangle) \\ &\quad + h_2(|\uparrow\downarrow, \uparrow\uparrow\rangle + |\downarrow\uparrow, \uparrow\uparrow\rangle + |\uparrow\downarrow, \downarrow\downarrow\rangle + |\downarrow\uparrow, \downarrow\downarrow\rangle)] \end{aligned} \tag{4.124}$$

This gives:

$$\langle \phi_0^{(I)} | 2S^x_{m\sigma} | \phi_0^{(I)} \rangle = -\frac{8}{U+J} h_m. \tag{4.125}$$

Analogously to the 1-band case to calculate the critical interaction one uses the self-consistency condition $h_m = 2\bar{\epsilon}_m \langle 2S^x_{m\sigma}\rangle$. This linearized condition determines the critical coupling U_c.

The self-consistency equations for each orbital then read:

$$h_m = 2\bar{\epsilon}_m \langle 2S^x_{1\sigma}\rangle = -\frac{16 h_m}{U+J}\bar{\epsilon}_m \implies U^m_c = -16\bar{\epsilon}_m - J. \tag{4.126}$$

This equations show two important results:

- In the half-filled Hubbard model with two identical bands (i.e., $\bar{\epsilon}_1 = \bar{\epsilon}_2$ here), at large J the two bands undergo a common Mott transition of which the U_c follows the Mott gap asymptotics (as described in Sect. 4.6.6). It is interesting however to notice that the intercept at $J = 0$ is the U_c for the Mott transition in the 1-band Hubbard model $U_c(N = 1) = -16\bar{\epsilon}$ (Sect. 4.5.5), so that the result for the half-filled 2-band model can be written:

$$U_c(N = 2, \text{large } J) = U_c(N = 1) - J. \tag{4.127}$$

- the equations for the two orbitals are *decoupled*, contrary to what happens at $J = 0$ (see (4.139)). This is due to the fact that when starting from the triplet ground state at finite J, the second order perturbation theory that leads to the linearized self-consistency equation necessarily implies two hopping processes (two slave-spin flips, in the context of the slave-spin mapping) in the same orbital, or one would end up in one of the states of the $E = 2J$ or $E = 4J$ groups, and thus outside the degenerate ground state manifold. This is not the case at $J = 0$ because all these states are degenerate within the ground state manifold of dimension 6. This is the low-energy *orbital decoupling* induced by Hund's coupling J, and it is materialized - in this large J limit[14] - in the fact that for $\bar{\epsilon}_1 \neq \bar{\epsilon}_2$ each band undergoes a separate Mott transition at a critical interaction proportional to the respective kinetic energy $\bar{\epsilon}_m$. Thus the Mott transition becomes *orbital-selective* (OSMT) and generates the phase diagram drawn with dashed lines in Fig. 4.13.

One last remark is in order. The approximate treatment of the spin-flip and pair-hopping terms of the Kanamori hamiltonian in SSMF might lead to consider a more physical result the one extracted from the Ising form of the interaction. Indeed this is the present standard of use in realistic calculations based on SSMF.

However the perturbative expansion performed in this section is justified if the transition is second order, since only in that case it makes sense to perform a calculation for vanishing h_m. This indeed happens for the Kanamori hamiltonian at large J (see Fig. 4.6 upper panel). In the case of the Ising form of Hund's coupling instead the transition is always first order at finite J (Fig. 4.6 lower panel), thus the perturbative treatment is not justified. It can nevertheless be performed (the treatment is strictly analogous to the one performed in this complement for the Kanamori case) and it yields the result:

$$U_c(N = 2, \text{large } J) = U_c(N = 1)/2 - J \qquad \text{("Ising" interaction)}, \tag{4.128}$$

(this smaller value for U_c compared to the value found for the Kanamori hamiltonian is indeed in line with the fact that the ground state degeneracy is brought down to a doublet in the Ising hamiltonian, instead than a triplet like in the Kanamori case).

---

[14] In this sense the limit is large-J: the states outside the lowest triplet are excluded altogether from the perturbation theory, which would happen rigorously only in the large-J limit.

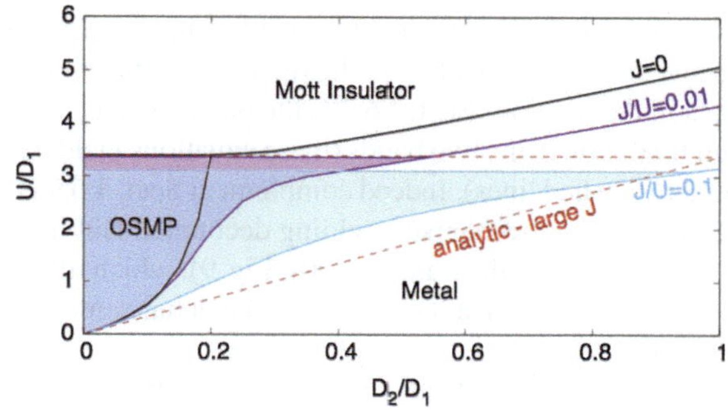

Fig. 4.13 Phase diagram of the 2-band Hubbard model with different bandwidths, in the plane bandwidth ratio (D_2/D_1) - interaction strength (U/D_1)), for several values of Hund's coupling J/U and for $J = 0$ (adapted from [21]). The analytic result showing how in the large-J limit the orbitals get decoupled is obtained in complement Sect. 4.6.9

This results signals that indeed a self-consistent solution exists for vanishing h_m, and it is thus a metastable one, which is pre-empted by the first-order transition (as studied in [39, 48]).

In any case these results collectively point to a $J = 0$ intercept of the large-J result for U_c of the order of the critical coupling for the single-band Hubbard model.

4.6.10 Hund's-Induced Orbital Decoupling

As suggested in complement Sect. 4.6.9, another striking effect of Hund's coupling is to favor orbital-selective Mott transitions (OSMT) [49]. This was first shown in [47] and studied in more detail in [21, 50]. The main outcome of these studies on the 2-band Hubbard model with different bandwidths (OSMT in this model has been intensively studied in the period 2004–2005, a quite complete list of references can be found in [46]) is summarized in Fig. 4.13. There, the colored areas represent - for different Hund's coupling strength J/U - an orbital-selective Mott phase (OSMP) arising in between a metal at low interaction strength U and a Mott insulator at high U.

A difference in bandwidth between the two bands (which have here semi-circular densities of states, respectively half-bandwidth D_1 and D_2, and do not hybridize between them) is necessary to trigger the OSMT, however at J $= 0$ a quite extreme bandwidth ratio $D_2/D_1 = 0.2$ is needed. Indeed when a band is more than five times wider than the other, orbital fluctuations that promote a unique Mott transition (as investigated in Appendix) cannot make up for the strong difference in bare kinetic energy between the two bands and an OSMT takes place.

Hund's coupling however is seen to promote the OSMT and quickly shift this critical bandwidth ratio to moderate values, up to a point in which, already for moderate Hund's coupling $J/U = 0.1$, a little difference in bandwidth is enough to separate the transitions between the bands. Complement Sect. 4.6.9, in which a large J is assumed, shows that the same mechanism leading to the reduction of the width of the Hubbard bands highlighted in the previous Sect. 4.6.7 is the one that leads to decoupled equations for the Mott transition in the two bands (these equations yield the phase diagram shown in Fig. 4.13 by dashed lines). Indeed complement Sect. 4.6.9 is the low-energy counterpart in SSMF of the mechanism yielding decoupled Hubbard bands pointed out by Koga et al. in [47] (and illustrated in [30], Fig. 9), which ultimately motivates also their reduction in width (since it counteracts the enhancement mechanism [43] mentioned in Sect. 4.6.7).

Indeed numerical evidence that the core of all OSMT is an *orbital decoupling* caused by Hund's J is provided in [33], where several multi-band models were analyzed with bands of different width or bands of identical width but finite crystal-field splitting. There, it is pointed out that Hund's coupling causes a change from a collective to an individual orbital behaviour in which then it is the individual bandstructure and/or population in each orbital (and correspondingly in each band, if there is no hybridization) that determines the correlation strength of the electrons with that orbital character. This results does not rely necessarily on the bands having different width, but only on the independence of the gap to charge excitations in each orbital. The correlation between charge excitations can be measured in these models and it is strongly suppressed in the vicinity of the half-filled Mott insulator [33, 51], where orbital-selective Mott transitions are realized.[15] Indeed the striking result that bands of identical width could undergo separate Mott transitions was indeed remarked in [53] and studied in [31, 36, 54]. In these cases it is the final orbital population that sets the correlation strength and when an orbital is individually half-filled it can become insulating independently of the others and an orbital-selective Mott phase can be realized.

In general, all the OSMT at finite J can be rationalized within the Hubbard criterion as a result of decoupled Hubbard bands [51]. This is illustrated in Fig. 4.14.

A finite inter-orbital hopping or local hybridization acts against Hund's coupling, as also does the crystal-field splitting. However as long as Hund's coupling remains the dominant scale determining the atomic multiplet and is able to quench orbital fluctuations, these can be treated as perturbations and the physics of orbital decoupling is robust against them. Hybridization however ultimately can win over Hund's coupling at low energy [55, 56] (however also competing non-local energy scales can prevail [57, 58]) turning the localized component of the OSMP into a very heavy delocalized state. Thus the OSMP, in presence of hybridization is typically turned into a phase with coexistence of strongly and weakly correlated electrons in the conduction bands. This is what happens in Fe-based superconductors, and it is the subject of the next section.

---

[15] Even for OSMT with 4 electrons in 3 bands like in [31] or 6 electrons in 5 bands [32, 52] - seemingly far from half-filling - the orbital-selective insulating gap is connected with the gap of the half-filled Mott insulator.

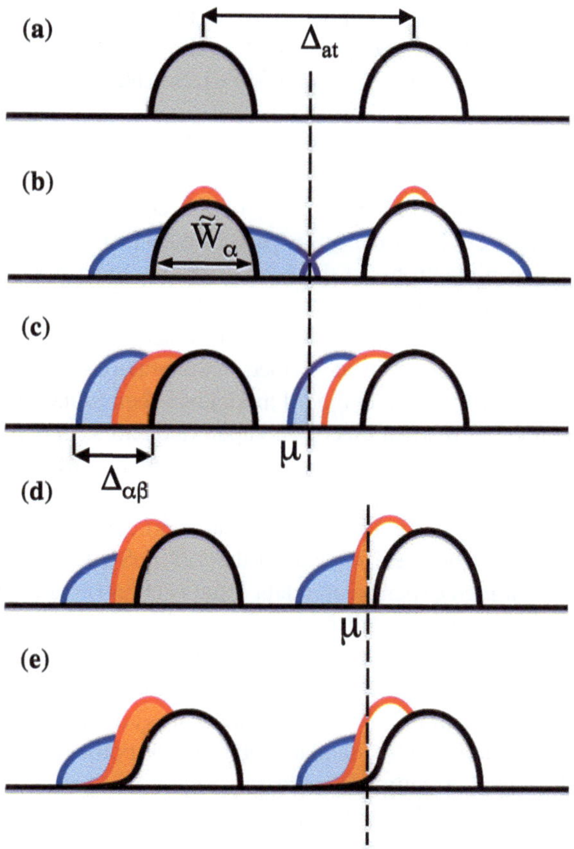

Fig. 4.14 Cartoon of a general mechanism for Hund's promoted orbital-selective Mott transitions and selective Mottness (see text). **a** Spectrum of a degenerate half-filled Hubbard model; **b** Orbital selective Mott phase triggered by a difference in bandwidths; **c** Orbital selective Mott phase triggered by a lifting of degeneracy of the orbital energies due to crystal-field splittings; **d** General situation with both the previous effects combined, OSM phase due to doping; **e** Onset of hybridization compared to **d**, transforming an orbital-selective Mott phase in a metallic phase with orbital-selective Mottness. From [51]

4.7 Slave-Spin Modeling of Fe-Based Superconductors

As anticipated in the introduction, SSMF has turned out very profitable in the investigation of the normal phase of Fe-based superconductors (FeSC). Indeed these are Fermi-liquid metals - apart from the phases where they turn superconductors - albeit with strongly renormalized band structures. Also the conduction bands are inherently multi-orbital in nature, and Hund's coupling is believed to play a major role. The large number of different materials belonging to this class and the large number of parameters determining the physics in these materials (interaction strength, pressure, doping, ...) make them a large playground, where an agile, albeit simplified, method for the realistic modeling of such phases like DFT+SSMF, can extract a lot of physical information and provide guidance in the understanding and predicting the physics of the presently known and unknown similar materials.

Most of the compounds from the different families have been studied with DFT+SSMF: LaFeAsO [23, 59], AFe_2As_2 (with A = Ba, K, Rb, Cs) [59–61], intercalated chalcogenides [62, 63] FeTeSe and the FeSe monolayer [64], $La_2O_3Fe_2Se_2$ [65].

The procedure for this realistic DFT+SSMF modeling is the following.[16] A tight binding fit (using a basis of localized wavefunctions, for example maximally localized Wannier functions [15]) of the conduction bands issued from the material-specific DFT simulations is performed. The local intra- and inter-orbital interactions between electrons on the orbitals of this local basis are included explicitly in this low-energy Hamiltonian, yielding thus a multi-orbital Hubbard model (4.96), which can be treated within SSMF as explained in the previous sections.[17]

All of the physics described in the previous sections and due to Hund's coupling in the multiorbital Hubbard model has been found to play an important role in these realistic investigations of Fe-superconductors. In particular an influence of the half-filled Mott insulator has been clearly found in these simulations.

Indeed one striking feature of these calculations is a steady increase of the normal-phase Sommerfeld coefficient (the coefficient of the low-temperature linear contribution to the specific heat) in the 122 FeSC family. This is illustrated in Fig. 4.15. The plotted experimental data (blue squares) concern $BaFe_2As_2$ throughout the full electron-doped (with partial Fe $\rightarrow$ Co substitution) and hole-doped (with Ba $\rightarrow$ K substitution) range, until the stoichiometric end-member KFe_2As_2, and further isovalent substitutions of the cations in this last compound giving stoichiometric $RbFe_2As_2$ and $CsFe_2As_2$. The measures are taken in high-purity single-crystals and the modeling with standard DFT (black points in Fig. 4.15) is clearly incapable to describe correctly this increase.

Some coupling to low-energy bosons (e.g. phonons or spin-fluctuations) or to critical fluctuations due to the proximity to a quantum critical point [60] can be invoked to try to explain this physics. However it appears simpler the explanation based on the fact that DFT modeling neglects local correlations, and these are particularly important for FeSC since Mott physics plays a major role.

Indeed the realistic modeling based on SSMF, which includes the local correlations as detailed in the previous sections and can describe the Mott transition,

[16]This is indeed completely analogous to the procedures used in the framework of DFT+DMFT calculations.

[17]If the low-energy tight-binding model includes local orbitals other than those in which the interaction is treated explicitly (e.g. typically if ligand-centered orbitals are also used in the local basis, for which the treatment of interaction is left at the DFT level, yielding a so-called "p-d model") then a relative energy shift between the correlated and uncorrelated orbitals has to be included. This "double-counting" term is of difficult assessment and many different recipes have been proposed [66–69]. The problem of having this, basically uncontrolled, extra parameter can be avoided altogether by limiting the low-energy tight-binding fit to the "correlated" orbitals (i.e., the five 3d Wannier functions centered on each Fe atom, in the present case of FeSC, yielding the so-called "d-model"). In this case the double counting term is absorbed in the chemical potential, which is fixed by the total electron population. The price to pay for this advantage is to have broader d-like Wannier functions, for which the Hubbard-like treatment is less justified. This is nevertheless the strategy that has been successfully followed thus far in all the published SSMF studies.

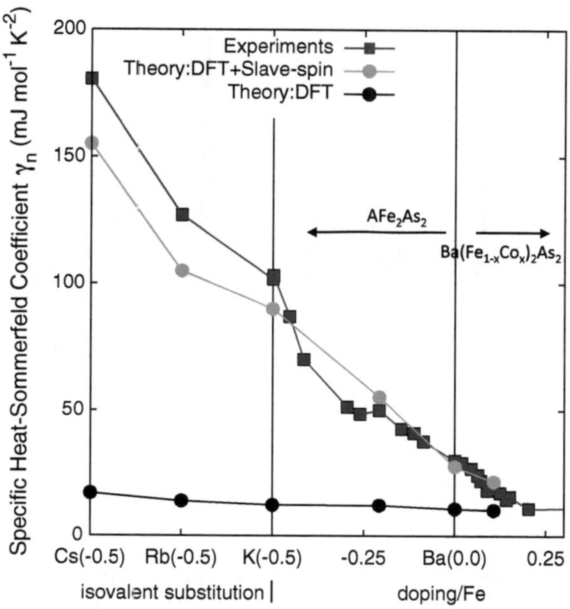

Fig. 4.15 Normal-phase Sommerfeld coefficient of the electronic specific heat in throughout the whole 122 family. Comparison between experiments (*blue squares*), DFT (*black circles*), DFT+Slave spins with a single set of interaction parameters (U = 2.7eV, J/U = 0.25) for all the compounds (*green circles*). Adapted from Hardy et al., [61] (color figure online)

predicts a Sommerfeld coefficient in very good agreement with experimental data (green points in Fig. 4.15), with only one set of interaction parameters throughout all the compounds (U = 2.7 eV, J/U = 0.25.[18])

It can be shown that in the investigated low-energy models for FeSC, for realistic values of the interaction parameters [70], a Mott insulator is always realized at half-filling (i.e., for an electronic density of 5 electrons per Fe atom, of which all the five 3d orbitals contribute predominantly to the conduction bands). At the density of 6 electrons/Fe, corresponding to that of the stoichiometric parent FeSC of all families (LaFeAsO, $BaFe_2As_2$, FeSe, etc.), no Mott transition is found until very high interaction strengths [23, 59, 64] (with the exception of the intercalated chalcogenides in the special Fe-vacancy-ordered phase [62] and of $La_2O_3Fe_2Se_2$, this being however a very different family, with very large crystal-field splitting strongly contrasting the effect of Hund's coupling on the Mott transitions [65]).

This can be shown in general, as indeed Mott physics depends very little on the details of the k-space band dispersions and mostly on k-integrated quantities, like the total kinetic energy of the DFT band structure. Indeed when normalizing the energy scale in a way to equate the bare kinetic energy of the realistic low-energy models for the various compounds to that of a simple 5-band Hubbard model with semicircular densities of states, one can show that the critical interaction strength for the half-filled Mott transitions are essentially coincident, and below the range of typical interaction values estimated for these compounds. The same phase diagram suggests that the

[18]This values depend on the type of Wannier orbitals used in the modeling of the conduction bands. These values are appropriate for the "d-model" scheme. See also footnote 17.

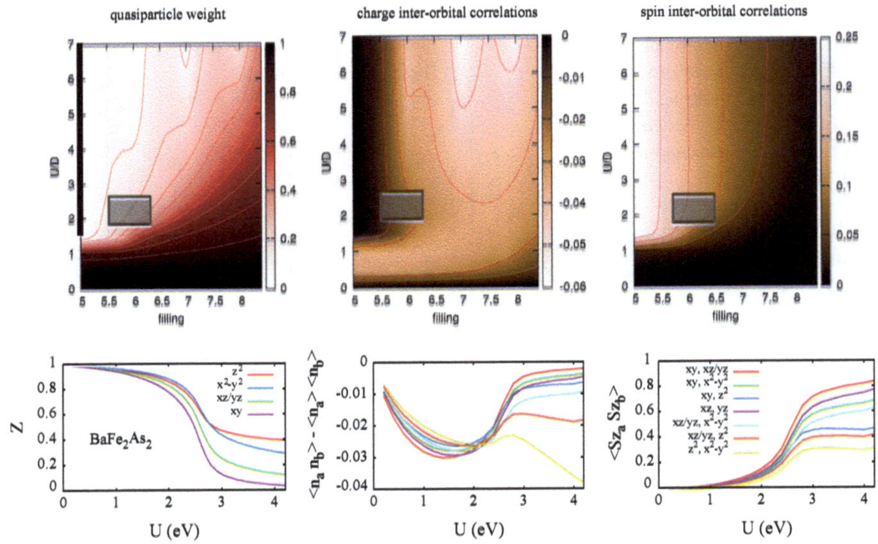

Fig. 4.16 *Upper panels* Quasiparticle weight (*left*), inter-orbital charge (*center*) and spin (*right*) correlations for the 5-orbital Hubbard model (with semi-circular DOS of half-bandwidth D) for J/U = 0.2, in the plane electron density (filling)/interaction strength (reproduced from [51]). The *black bar* on the *left graph* marks the Mott insulating phase at half-filling. The *gray box* locates the relevant zone for FeSC (see text). *Lower panels*. Ab-initio modeling for $BaFe_2As_2$ within SSMF, from [59] (Supplementary material). Same quantities as above at fixed filling n = 6, as a function of interaction strength for J/U = 0.25. The estimated value of U is 2.7–2.8 eV [70] (see footnote 18)

electron densities relevant for FeSC (5.5 ÷ 6.3 electrons/Fe) fall in general in the zone of influence of this half-filled Mott Insulator. Both these relevant ranges in the interaction/density plane are indicated by the grey box in Fig. 4.16 (upper panels).

Thus the experimentally measured strong enhancement of the Sommerfeld coefficient is simply explained with the materials approaching a Mott insulator which would be realized at half filling.[19] Indeed the Sommerfeld coefficient [6, 13] is

$$\gamma = \frac{\pi^2 k_B^2}{3} N^*(\epsilon_F) \tag{4.129}$$

with $N^*(\epsilon_F)$ the total quasiparticle density of states at the Fermi energy ϵ_F. In our SSMF framework this reduces to the DFT density of states at ϵ_F for $U = J = 0$, while its value is enhanced at finite interaction strength. This is an outcome of the quasiparticle mass enhancement induced by the interactions (in our method the

[19]The realization of a Mott insulator for half-filled FeSC is hard to test directly. Indeed the end member of the family KFe_2As_2 has a density of 5.5 electrons/Fe and no further hole-doped material has been synthesized in the family. A similar material with 5 electrons in the 5 conduction bands is $BaMn_2As_2$ however, which is found insulating [71] indeed, albeit the nature of this insulating state is still being clarified [72, 73].

renormalization of the hopping integrals), which produces bands which are less dispersive, thus denser in energy.

But besides the overall enhancement of the quasiparticle masses mainly set by the distance in doping from the half-filled Mott insulator, the influence of the proximity to this Mott phase in FeSC shows up in the several other aspects. These are seen in all realistic DFT+SSMF calculations (Fig. 4.16, lower panels) and are the main outcomes of the Hund's physics illustrated in the previous section, thus already present at the model level (Fig. 4.16, upper panels). Indeed ferromagnetic spin correlations between electrons in different orbitals are favored by Hund's coupling, thus leading to the formation of a large local moment in the paramagnetic metallic phase (this aspect received recently support from X-ray emission experiments, [74]). Also inter-orbital charge correlations are suppressed, due to the Hund's induced orbital-decoupling (the mechanism detailed in Sect. 4.6.10). Overall these features develop after a rapid crossover as a function of the interaction strength (clearly identifiable in the lower panels of Fig. 4.16 by the sharp changes around $U = 2.5 \div 2.8$ eV in all the quantities calculated for $BaFe_2As_2$, and by the accumulation of the contour lines in the model calculation in the upper panels), defining a broad frontier departing from the half-filled Mott transition [42, 75].

The successful denomination of "Hund's metal" [76] might be used for the zone beyond this frontier, where the Hund's coupling is essential in shaping the aforementioned phenomenology of the doped half-filled Mott insulator. This is more of a matter of taste however, for the present lack of a precise definition of a Hund's metal.

Another remarkable feature of this zone is that for temperatures higher than the Fermi-liquid coherence scale, the metal develops a finite spin-spin correlation at long times (and a sublinear frequency dependence of the self-energy at low frequency) which has been dubbed "spin-freezing" in the model context [77] and has been found in realistic simulations for FeSC [78, 79]. The coherence temperature is strongly suppressed, when approaching the half-filled Mott insulator (see discussion in [30] and references therein).

Predictably from the model studies (see Sect. 4.6.10), the features of the realistic band structure that break the orbital symmetry (most notably the crystal-field splitting of the orbital energies) induce an orbital differentiation of the correlation strength, as is clear from the lower-left panel in Fig. 4.16 in which one sees that for $BaFe_2As_2$, in the strongly correlated region just after the crossover, mass enhancements range from 2 to 10 depending on the orbital. This kind of orbital differentiation has been found in all the abovementioned simulations of FeSC within DFT+SSMF, and within many other methods: DMFT [75, 78, 80–85], variational Montecarlo [86, 87], Slave-spins [23, 62], Hartree–Fock mean-field [88], fluctuation-exchange approximation [89], Gutzwiller approximation [90].

The actual degree of orbital differentiation in the material depend on the realistic value of the interaction strength, which is estimated U = 2.7–2.8 eV in $BaFe_2As_2$ [70],[20] yielding mass enhancements ranging between 1.5 and 3, consistently with the experimental estimate.

A very detailed comparison between theoretical predictions and experimental estimates of the mass enhancement was done in [59], clearly establishing the orbital selectivity of electronic correlations in the whole phase diagram of the 122 family and assessing the degree of selectivity. Indeed (as shown in the upper panel of Fig. 4.17) a substantial discrepancy among the various experimental data quantifying the correlation strength is found, when trying to uniquely define if electrons are strongly *or* weakly correlated in these materials. As already mentioned the Sommerfeld coefficient raises uniformly, indicating a strong mass enhancement, from the electron-overdoped (where the electrons are essentially non correlated) to the stoichiometric compounds (where the mass enhancement is ~3), all the way to the hole-overdoped family end-members (where the mass enhancement is ~9). On the other hand other probes like low-frequency optical conductivity (of which the integrated spectral weight is lowered by correlations) or ARPES (that can directly measure the reduction in band dispersion due to the mass enhancement) report much more moderate correlation strengths over the whole phase diagram (topping at ~3 for KFe_2As_2).

However this discrepancy is readily interpreted as a *coexistence of weakly and strongly correlated electrons*. Indeed when the mass enhancement is not equal for all the electrons the different experimental quantities can be renormalized in very different ways. For instance the heaviest electrons tend to strongly enhance the Sommerfeld coefficient, which is a sum of contributions from each orbital character, each weighed by the respective mass enhancement, to the total density of states. On the other hand the Drude weight (the integrated low-frequency spectral weight of the optical conductivity) is a sum of contributions from each orbital each weighed by its *inverse* mass enhancement. Thus the contribution from the most renormalized electrons tends to be suppressed, and the renormalization of the lighter electrons will tend to dominate the renormalization of the total weight. Analogously, the band-structure, which is an intricate manifold of conduction bands that can be seen as the hybridization due to inter-orbital hoppings of five Fe 3d bands (one for each orbital), will still disperse as the more dispersive of these bands, even if selective correlations strongly reduce the dispersion corresponding to some orbitals.

Thus, naturally, estimates of the correlation strength from optics or ARPES whole bandstructures will yield much smaller values compared to specific heat estimates, if weakly and strongly correlated electrons coexist in the conduction bands.

ARPES and quantum oscillations can however also probe the renormalization of each band separately. Indeed they yield values for the mass enhancement in the different sheets of the Fermi surface (see these values overlaid to Fig. 4.17 in [59])

[20] A sizable error bar is actually to be considered around these c-RPA estimated values, U in particular. A small 5 % adjustment in LaFeAsO for instance was found necessary in [87], but the error on U can in principle be substantially higher.

Fig. 4.17 Mass enhancement for doped $BaFe_2As_2$ and KFe_2As_2. Experiments (*upper panel*, experimental references in [59] and supplementary material) and theory within SSMF (U = 2.7 eV, J/U = 0.25). *Triangles* (stoichiometric) and *solid lines* (doped) correspond to calculations performed using the DFT bandstructure for $BaFe_2As_2$, while the *squares* correspond to the calculation for stoichiometric KFe_2As_2. The different colors indicate electrons of different orbital character (see legend)

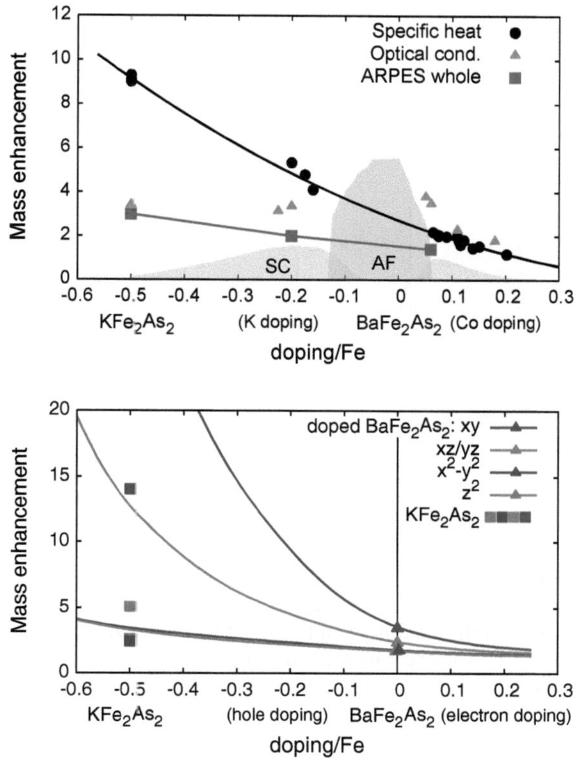

which are concentrated around $\sim 2 \div 3$ in the electron doped side of the phase diagram, spread progressively with decreasing filling, and range between 2 and 20 for KFe_2As_2. This is in agreement with the theoretical estimates (reported in the lower panel of Fig. 4.17).

As predicted by the theory (already in [31–33] and then in most of the aforementioned works) the strongest renormalization always happen for the electrons of the xy orbital character after being theoretically predicted in the stronger correlations in xy electrons.[21] The main reason for the xy orbital being the most correlated of all is its final population, which is the closest to half-filling among the five orbitals. Indeed in the framework of Hund's-induced orbital-decoupling detailed in Sect. 4.6.10, each orbital is distinctly influenced by Mott physics ("selective Mottness" [59]). Indeed in this regime the enhancement of the correlation among electrons in each orbital depends on the proximity to the Mott state that each orbital can reach separately once individually half-filled. A more thorough discussion on this mechanism and a

[21] One should mention that in some works an alternative convention is used and the orbital label xy is switched with $x^2 - y^2$. It is just a question of nomenclature however, the most correlated orbital being the planar one of t_{2g} symmetry, that we call xy following the most used convention.

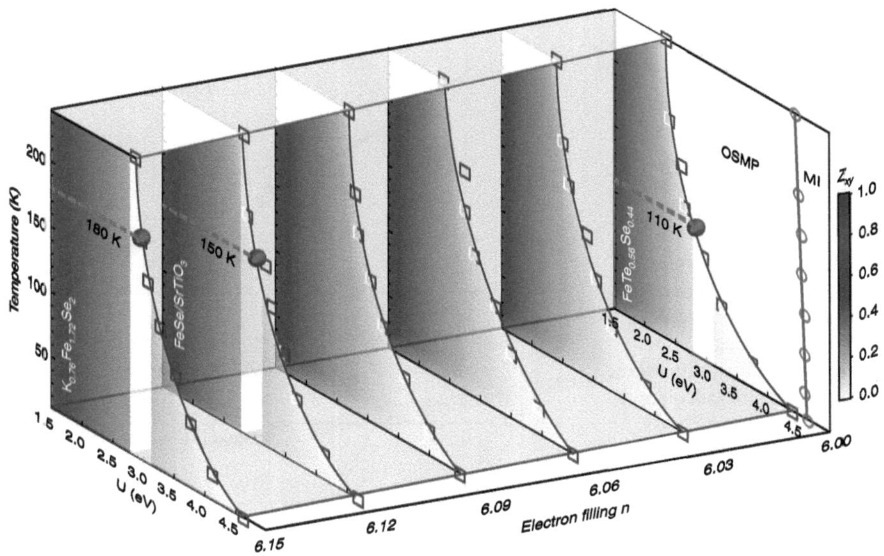

Fig. 4.18 Coherence scale (*blue lines*) for the xy electrons in several Fe-chalcogenides within SSMF. Hund's coupling is kept to a fixed value throughout these calculations, thus causing the Mott transition for 6 electrons in this 5-orbital model to happen at a lower U_c than in the modeling at fixed J/U. The experimentally determined values are marked by *pink lines*. From [64] (color figure online)

specific analysis of the stronger correlations in the xy orbital in FeSC can be found in the last sections and appendix of [51].

The stronger correlations of the xy electrons imply a lower coherence temperature for the corresponding quasiparticles. When this temperature is crossed indeed a quick depletion of the spectral weight associated to xy quasiparticles has been directly observed in ARPES, in [63, 64] in most of the Iron chalcogenides. This was successfully modeled within SSMF as shown in Fig. 4.18.

Another useful application of SSMF has been the determination of the contribution of quasiparticle interband transitions to the infrared optical conductivity of the 122 family [91]. Indeed the renormalization of the bandstructure induces a shift of these transitions to lower frequency compared to the DFT estimate, and in this work it is predicted that a substantial contribution from these transitions extends down to the whole range of the mid-infrared.

4.8 Concluding Remarks

In this chapter we gave a pedagogical introduction to the slave-spin technique. After an introduction on the physics of strongly correlated materials, and of the Fermi-liquid metallic phases, alongside with the details of the technique we have seen the main aspects of the influence of Mott and Hund physics in the correlated metallic phases. We have finally applied these known concepts issued of both old and recent research to the physics of the recently discovered Fe-based superconductors. Some new concepts are also introduced here, that are useful to complete our view on the normal phases of these superconductors: the shrinking of the Hubbard bands of the half-filled multi-orbital Mott insulator and the low-energy description of the "orbital-decoupling", both induced by a sizable Hund's coupling.

Obviously this manuscript is a path through many topics, bridging several theoretical and experimental areas in the field of strongly correlated materials, that is exquisitely the result of the taste of the authors, and no claim of completeness could be associated to it, in any of the respective areas. Thus, for example, it cannot be a complete treatment of the physics of Iron-based superconductors, far from it actually, which receive a much more complete and equilibrate treatment in many of the recent and less recent reviews available in the literature (a non-exhaustive list being the following [92–96] and references therein). Nor it is a complete treatment of Mott and Hund's physics, of which the interested reader can deepen his knowledge in another series of available reviews (a very incomplete list being the following [17, 30, 45, 97, 98]). Finally, even on the slave-spin technique the focus is put primarily on the flavor of this technique used by the authors, and the reader is redirected to [23] for the main alternative - albeit very similar both in the spirit and in the results on FeSC - take, besides the various publications mentioned in Sect. 4.7.

However we believe to have given our best possible intersection of the three previous subjects, and in this sense this manuscript probably represent a rather thorough coverage.

It is worth mentioning two natural extensions of the SSMF that have not been worked out yet, still they are very much needed: the formalism for symmetry-broken (magnetic, superconducting etc.) phases, and the rotationally invariant formulation, capable to treat the Kanamori–Hund interaction on the same level of approximation than the Ising form.

Appendix: Mott Transition in the 2-Orbital Hubbard Model with Different Bandwidths at $J = 0$

Here we show analytically that in the half-filled degenerate 2-band Hubbard model with different bandwidths (due to different diagonal hopping integrals t_1 and t_2), the Mott transition at $J = 0$ remains common between the two bands even when $t_1 \neq t_2$, until quite large bandwidth rations. However, in accordance with the numerical evidence the largest bandwidth ratio where this is possible is $t_1/t_2 = 5$. For larger ratios the U_c for the Mott transition differs between the bands, and one opens the $J = 0$ orbital-selective Mott phase seen in Fig. 4.13.

As done previously we develop the multi-orbital slave-spin equation (4.100) around the insulating solution $h_{im\sigma} = 0$ where $H_s = H_{at}$ with $J = 0$:

$$H_{at} = \frac{U}{2}\left(\sum_{m\sigma} S^z_{m\sigma}\right)^2, \tag{4.130}$$

of which we know the spectrum. This is:

$$\begin{cases} |\uparrow\uparrow,\uparrow\uparrow\rangle |\downarrow\downarrow,\downarrow\downarrow\rangle & E = 2U \\ |\uparrow\uparrow,\uparrow\downarrow\rangle |\uparrow\uparrow,\downarrow\uparrow\rangle |\downarrow\downarrow,\uparrow\downarrow\rangle |\downarrow\downarrow,\downarrow\uparrow\rangle & \\ |\uparrow\downarrow,\uparrow\uparrow\rangle |\downarrow\uparrow,\uparrow\uparrow\rangle |\uparrow\downarrow,\downarrow\downarrow\rangle |\downarrow\uparrow,\downarrow\downarrow\rangle & E = \frac{U}{2} \\ |\uparrow\downarrow,\uparrow\downarrow\rangle |\uparrow\downarrow,\downarrow\uparrow\rangle |\downarrow\uparrow,\uparrow\downarrow\rangle |\downarrow\uparrow,\downarrow\uparrow\rangle & \\ |\uparrow\uparrow,\downarrow\downarrow\rangle |\downarrow\downarrow,\uparrow\uparrow\rangle & E = 0 \end{cases} \tag{4.131}$$

We again calculate as a perturbation, the effect of the "kinetic" part of the slave-spin hamiltonian:

$$H_{pert} = \sum_{m\sigma} h_m 2S^x_{m\sigma}. \tag{4.132}$$

where $O_{im\sigma} = 2S^x_{m\sigma}$ because both orbitals are half-filled and thus the gauge $c = 1$ ((4.61) for $n_m = 1/2$), and we call $h_m \equiv h^*_{im\sigma} + h_{im\sigma}$. Also $\lambda_m = 0$ at half-filling because of particle-hole symmetry. Notice that here we let h_1 and h_2 differ.

We perform as in the other cases a perturbative expansion in h_1 and h_2, which are small near the metal-insulator transition, if this is second-order.

In this $J = 0$ case the ground state has degeneracy 6 (the subset above with $E = 0$). Furthermore, as before, the perturbation H_{pert} does not have any nonzero elements in the low-energy subspace one has to use second-order perturbation theory to find the right combination of states, in the ground state manifold, to which the perturbed state tends for $h_m \to 0$. This is the ground state of the matrix $H' \equiv H_{pert}(E_0 - H_{at})^{-1} H_{pert}$ in the degenerate subspace. In this case, since H_{pert} couples the $E = 0$ sector only with the $E = \frac{U}{2}$ sector, one can rewrite $H' = -\frac{2}{U} H^2_{pert}$ (where we have used the fact that $E_0 - H_{at}$ is diagonal and $E_0 = 0$).

By applying H' to one of the states with orbital polarization (i.e., $|\uparrow\uparrow,\downarrow\downarrow\rangle$ and $|\downarrow\downarrow,\uparrow\uparrow\rangle$ that corresponds to the original states in which the two electrons occupy both the same orbital), one finds that both are only coupled with one combination of the four non-polarized states, which is

$$|+,+\rangle = \frac{1}{2}(|\uparrow\downarrow,\uparrow\downarrow\rangle + |\uparrow\downarrow,\downarrow\uparrow\rangle + |\downarrow\uparrow,\uparrow\downarrow\rangle + |\downarrow\uparrow,\downarrow\uparrow\rangle) \tag{4.133}$$

In detail:

$$\begin{aligned} H'|\uparrow\uparrow,\downarrow\downarrow\rangle &= \left[2(h_1^2+h_2^2)|\uparrow\uparrow,\downarrow\downarrow\rangle + 2h_1^2\,\downarrow\downarrow,\downarrow\downarrow\rangle + 2h_2^2|\uparrow\uparrow,\uparrow\uparrow\rangle + 2h_1h_2(|++\rangle)\right] \\ H'|\downarrow\downarrow,\uparrow\uparrow\rangle &= [\quad \text{“} \quad h_1 \longleftrightarrow h_2 \quad \text{”} \quad] \end{aligned} \tag{4.134}$$

Indeed V^2 has the effect of flipping any two spins (including twice the same), so that any state of the $E = 0$ manifold is changed, when acted upon by V^2, into itself plus four of the others states of the manifold (only one is left out, since it needs 4 flips to be reached), plus two states of the $E = 2U$ manifold.

This suggest to pick the following three other orthogonal states[22] as basis states for writing H':

$$
\begin{aligned}
|+,-\rangle &= \frac{1}{2}(|\uparrow\downarrow,\uparrow\downarrow\rangle - |\uparrow\downarrow,\downarrow\uparrow\rangle + |\downarrow\uparrow,\uparrow\downarrow\rangle - |\downarrow\uparrow,\downarrow\uparrow\rangle)\\
|-,+\rangle &= \frac{1}{2}(|\uparrow\downarrow,\uparrow\downarrow\rangle + |\uparrow\downarrow,\downarrow\uparrow\rangle - |\downarrow\uparrow,\uparrow\downarrow\rangle - |\downarrow\uparrow,\downarrow\uparrow\rangle)\\
|-,-\rangle &= \frac{1}{2}(|\uparrow\downarrow,\uparrow\downarrow\rangle - |\uparrow\downarrow,\downarrow\uparrow\rangle - |\downarrow\uparrow,\uparrow\downarrow\rangle + |\downarrow\uparrow,\downarrow\uparrow\rangle)
\end{aligned}
\tag{4.135}
$$

It is straightforward to verify that the restriction of H' to the $E = 0$ manifold in this basis is given by the following 6×6 matrix:

$$
H' = -\frac{2}{U}\begin{bmatrix}
2h_1^2+2h_2^2 & 0 & 4h_1h_2 & 0 & 0 & 0\\
0 & 2h_1^2+2h_2^2 & 4h_1h_2 & 0 & 0 & 0\\
4h_1h_2 & 4h_1h_2 & 4h_1^2+4h_2^2 & 0 & 0 & 0\\
0 & 0 & 0 & 4h_1^2 & 0 & 0\\
0 & 0 & 0 & 0 & 4h_2^2 & 0\\
0 & 0 & 0 & 0 & 0 & 0
\end{bmatrix}
\tag{4.136}
$$

The diagonalization of this matrix yields the ground state energy at the first order in the perturbation i.e., $E_0^I = (-6(h_1^2 + h_2^2) - 2\sqrt{(h_1^2 + h_2^2)^2 + 32h_1^2h_2^2})/U$ and the correct zeroth order ground state, that is

$$
|\phi_0\rangle = X(|\uparrow\uparrow,\downarrow\downarrow\rangle + |\downarrow\downarrow,\uparrow\uparrow\rangle) + Y|+,+\rangle \tag{4.137}
$$

with $X = \frac{4h_1h_2}{\sqrt{\mathcal{N}}}$, $Y = \frac{h_1^2+h_2^2+\sqrt{(h_1^2+h_2^2)^2+32h_1^2h_2^2}}{\sqrt{\mathcal{N}}}$, and $\mathcal{N} = 64h_1^2h_2^2 + 2(h_1^2 + h_2^2)^2 + 2(h_1^2 + h_2^2)\sqrt{(h_1^2 + h_2^2)^2 + 32h_1^2h_2^2}$.

To first order in the perturbation the ground state becomes[23]:

[22]This notation is easily understood as follows: the state for the two-orbital system is obtained by tensor product of two one-orbital states, which are chosen either the following triplet state $|+\rangle \equiv (|\uparrow\downarrow\rangle + |\downarrow\uparrow\rangle)/\sqrt{2}$, or the singlet state $|-\rangle \equiv (|\uparrow\downarrow\rangle - |\downarrow\uparrow\rangle)/\sqrt{2}$. Thus for example the state $|+,-\rangle$ is the tensor product of a triplet state in orbital 1 and the singlet state in orbital two, yielding the first formula in (4.135).

[23]Here, as specified previously, $|+,\downarrow\downarrow\rangle$ is intended as the tensor product of the triplet state $|+\rangle \equiv (|\uparrow\downarrow\rangle + |\downarrow\uparrow\rangle)/\sqrt{2}$ for orbital 1 with the state $|\downarrow\downarrow\rangle$ for orbital 2, i.e., $|+,\downarrow\downarrow\rangle \equiv (|\uparrow\downarrow,\downarrow\downarrow\rangle + |\downarrow\uparrow,\downarrow\downarrow\rangle)/\sqrt{2}$.

$$
\begin{aligned}
|\phi_0^{(I)}\rangle &= |\phi_0\rangle + \sum_{|s\rangle \neq |\phi_0\rangle} \frac{\langle s|H_{pert}|\phi_0\rangle}{E_0 - E_s}|s\rangle \\
&= |\phi_0\rangle - \frac{2}{U}X[\sqrt{2}h_1(|+,\downarrow\downarrow\rangle + |+,\uparrow\uparrow\rangle) + \sqrt{2}h_2(|\uparrow\uparrow,+\rangle + |\downarrow\downarrow,+\rangle)] \\
&\quad -\frac{2}{U}Y[\sqrt{2}h_1(|\uparrow\uparrow,+\rangle + |\downarrow\downarrow,+\rangle) + \sqrt{2}h_2(|+,\downarrow\downarrow\rangle + |+,\uparrow\uparrow\rangle)] \qquad (4.138)
\end{aligned}
$$

We now use the self-consistency conditions $h_m = 2\bar{\epsilon}_m \langle \phi_0^{(I)}|2S^x_{m\sigma}|\phi_0^{(I)}\rangle$ to obtain, as in the complements Sects. 4.5.5 and 4.6.4 the linearized equations for the h_m which determine the critical coupling for the Mott transition. By calculating $\langle 2S^x_{m\sigma}\rangle$ for both orbitals we get to the following coupled equations:

$$
\begin{aligned}
h_1 &= 2\bar{\epsilon}_1(-\frac{4}{U})\left[(2X^2+2Y^2)h_1 + 4XYh_2\right] \\
h_2 &= 2\bar{\epsilon}_2(-\frac{4}{U})\left[(2X^2+2Y^2)h_2 + 4XYh_1\right] \qquad (4.139)
\end{aligned}
$$

These equations as usual determine, for $(h_1, h_2) \to 0$, the U for the Mott transition, as long as h_1 and h_2 vanish together. But the equations are two because, since h_1 and h_2 are determined selfconsistently they vanish with a given ratio $\alpha \equiv h_2/h_1$, which is fixed by the asymmetry between the bands, namely by $\bar{\epsilon}_2/\bar{\epsilon}_1$.

By substituting $h_2 = \alpha h_1$ in (4.139) and in the expressions for X and Y, and simplifying h_1 everywhere one gets two equations determining U_c and α for given $\bar{\epsilon}_1$ and $\bar{\epsilon}_2$. U can be eliminated to get the equation determining implicitly α for each $\bar{\epsilon}_2/\bar{\epsilon}_1$ that reads:

$$
\frac{\bar{\epsilon}_2}{\bar{\epsilon}_1} = \frac{X^2+Y^2+2\alpha XY}{X^2+Y^2+\frac{2}{\alpha}XY} \qquad (4.140)
$$

with $X = \frac{4\alpha}{\sqrt{\mathcal{N}_\alpha}}$, $Y = \frac{1+\alpha^2+\sqrt{(1+\alpha^2)^2+32\alpha^2}}{\sqrt{\mathcal{N}_\alpha}}$, and $\mathcal{N}_\alpha = 64\alpha^2 + 2(1+\alpha^2)^2 + 2(1+\alpha^2)\sqrt{(1+\alpha^2)^2+32\alpha^2}$. This formula gives the dependence shown in Fig. 4.19. This shows that a common Mott transition within the two bands can be found for any bandwidth ratio between 1 and 0.2, which is the limiting ratio. This can also be calculated by expanding for small α. One has $X^2+Y^2 \simeq 1$ and $XY \simeq 2\alpha$ at the leading order, giving:

$$
U_c = -16\bar{\epsilon}_1, \qquad \bar{\epsilon}_2/\bar{\epsilon}_1 = 0.2. \qquad (4.141)
$$

which locates the point where the three transition lines merge as found in the numerical calculation shown in Fig. 4.13 for $J = 0$ [21, 50].

For $\bar{\epsilon}_2/\bar{\epsilon}_1 = 1$ instead, one has $\alpha = 1$ by symmetry and recovers the result of the complement Sect. 4.6.4 for the U_c in the degenerate 2-band Hubbard model at $J = 0$ by evaluating $X = 1/\sqrt{6}$, $Y = 2/\sqrt{6}$, which yield indeed $U_c = -24\bar{\epsilon}_1 = -24\bar{\epsilon}_2$.

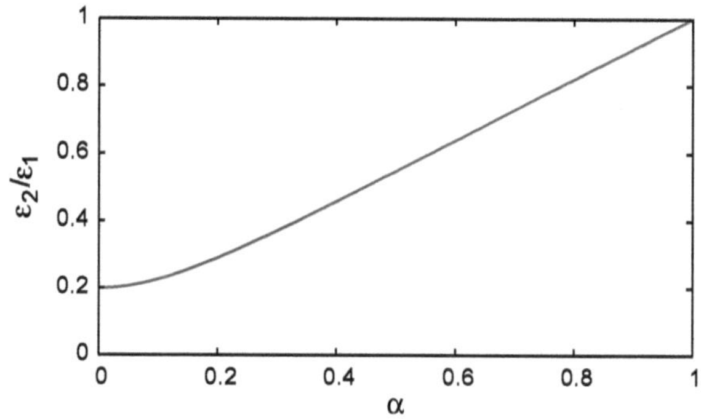

Fig. 4.19 The relation between $\alpha \equiv h_2/h_1$ and $\bar{\epsilon}_2/\bar{\epsilon}_1$ implied in formula (4.140)

References

1. Y. Kamihara, H. Hiramatsu, M. Hirano, R. Kawamura, H. Yanagi, T. Kamiya, H. Hosono, J. Am. Chem. Soc. **128**, 10012 (2006)
2. J.G. Bednorz, K.A. Müller, Zeitschrift fur Physik B Condensed Matter **64**, 189–193 (1986)
3. P. Anderson, Science **235**, 1196 (2007)
4. I.I. Mazin, D.J. Singh, M.D. Johannes, M.H. Du, Phys. Rev. Lett. **101**, 057003 (2008)
5. Q. Si, E. Abrahams, Phys. Rev. Lett. **101**, 076401 (2008)
6. N. Ashcroft, N. Mermin, *Solid State Physics* (Saunders College, Philadelphia, 1976)
7. W. Kohn, L. Sham, Phys. Rev. **140**, A1133 (1965)
8. N.F. Mott, Proc. Phys. Soc. (London) **A62**, 416–422 (1949)
9. J.H. de Boer, E.J.W. Verwey, Proc. Phys. Soc. **49**, 59 (1937)
10. J. Zaanen, G.A. Sawatzky, J.W. Allen, Phys. Rev. Lett. **55**, 418 (1985)
11. L. Landau, Sov. Phys. JETP **3**, 920–925 (1957)
12. L. Landau, Sov. Phys. JETP **8**, 70–74 (1959)
13. P. Nozières, *Theory of Interacting Fermi systems* (Westview Press, Boulder, 1997)
14. G.R. Stewart, Rev. Mod. Phys. **56**, 755–787 (1984)
15. N. Marzari, A.A. Mostofi, J.R. Yates, I. Souza, D. Vanderbilt, Rev. Mod. Phys. **84**, 1419–1475 (2012)
16. E. Dagotto, Rev. Mod. Phys. **66**, 763–840 (1994)
17. M. Imada, A. Fujimori, Y. Tokura, Rev. Mod. Phys. **70**, 1039–1263 (1998)
18. P. Lee, N. Nagaosa, X.G. Wen, Rev. Mod. Phys. **78**, 17 (2006)
19. W. Metzner, M. Salmhofer, C. Honerkamp, V. Meden, K. Schönhammer, Rev. Mod. Phys. **84**, 299–352 (2012)
20. G. Kotliar, A. Ruckenstein, Phys. Rev. Lett. **57**, 1362 (1986)
21. de' Medici L, Georges A, Biermann S., Phys. Rev. B **72**, 205124 (2005)
22. SRHassan, de' Medici L., Phys. Rev. B **81**, 035106 (2010)
23. R. Yu, Q. Si, Phys. Rev. B **86**, 085104 (2012)
24. A. Rüegg, S.D. Huber, M. Sigrist, Phys. Rev. B **81**, 155118 (2010)
25. M. Schiró, M. Fabrizio, Phys. Rev. B **83**, 165105 (2011)
26. M. Mardani, M.S. Vaezi, A. Vaezi, (2011), arXiv:1111.5980
27. Y. Zhong, H.G. Luo, (2012), arXiv:1204.4777
28. A.B. Georgescu, S. Ismail-Beigi, Phys. Rev. B **92**, 235117 (2015)
29. L. de' Medici, PhD Thesis, Ecole Polytechnique and Université de Paris-Sud (2006)
30. A. Georges, L. de' Medici, J. Mravlje, Annu. Rev. Condens. Matter Phys. **4**, 137–178 (2013)
31. de' Medici L, SRHassan, Capone M, Dai X., Phys. Rev. Lett. **102**, 126401 (2009)
32. L. de' Medici, S. Hassan, M. Capone, J. Supercond. N. Mag. **22**, 535 (2009)
33. de' Medici L., Phys. Rev. B **83**, 205112 (2011)

34. P. Lu, Phys. Rev. B **49**, 5687 (1994)
35. M.J. Rozenberg, Phys. Rev. B **55**, R4855 (1997)
36. P. Werner, E. Gull, A. Millis, Phys. Rev. B **79**, 115119 (2009)
37. S. Florens, A. Georges, G. Kotliar, O. Parcollet, Phys. Rev. B **66**, 205102 (2002)
38. F. Lechermann, A. Georges, G. Kotliar, O. Parcollet, Phys. Rev. B **76**, 155102 (2007)
39. Y. Ono, M. Potthoff, R. Bulla, Phys. Rev. B **67**, 035119 (2003)
40. J. Hubbard, Proc. R. Soc. Lond. Ser. A. Math. Phys. Sci. **276**, 238–257 (1963)
41. de' Medici L, Mravlje J, Georges A., Phys. Rev. Lett. **107**, 256401 (2011)
42. L. Fanfarillo, E. Bascones, Phys. Rev. B **92**, 075136 (2015)
43. O. Gunnarsson, E. Koch, R. Martin, Phys. Rev. B **54**, R11026 (1996)
44. O. Gunnarsson, E. Koch, R. Martin, Phys. Rev. B **56**, 1146 (1997)
45. A. Georges, G. Kotliar, W. Krauth, M.J. Rozenberg, Rev. Mod. Phys. **68**, 13–125 (1996)
46. K. Inaba, A. Koga, Phys. Rev. B **73**, 155106 (2006)
47. A. Koga, NKawakami, Rice T, Sigrist M., Phys. Rev. Lett. **92**, 216402 (2004)
48. N. Lanatà, H.U.R. Strand, X. Dai, B. Hellsing, Phys. Rev. B **85**, 035133 (2012)
49. V. Anisimov, I. Nekrasov, D. Kondakov, T. Rice, M. Sigrist, Eur. Phys. J. B **25**, 191 (2002)
50. M. Ferrero, F. Becca, M. Fabrizio, M. Capone, Phys. Rev. B **72**, 205126 (2005)
51. L. de' Medici, *Weak and Strong Correlations in Iron pnictides*, Springer Series in Materials Science (Springer, Berlin, 2015), pp. 409–441
52. A.M. Läuchli, P. Werner, Phys. Rev. B **80**, 235117 (2009)
53. P. Werner, A.J. Millis, Phys. Rev. Lett. **99**, 126405 (2007)
54. L. Huang, L. Du, X. Dai, Phys. Rev. B **86**, 035150 (2012)
55. E. Pavarini, L.C. Andreani, Phys. Rev. Lett. **77**, 2762–2765 (1996)
56. E.A. Winograd, de' Medici L., Phys. Rev. B **89**, 085127 (2014)
57. C. Pepin, Phys. Rev. Lett. **98**, 206401 (2007)
58. L. De Leo, M. Civelli, G. Kotliar, Phys. Rev. B **77**, 075107 (2008)
59. de' Medici L, Giovannetti G, Capone M., Phys. Rev. Lett. **112**, 177001 (2014)
60. F. Eilers, K. Grube, D.A. Zocco, T. Wolf, M. Merz, P. Schweiss, R. Heid, R. Eder, R. Yu, J.X. Zhu, Q. Si, T. Shibauchi, Löhneysen Hv, Phys. Rev. Lett. **116**, 237003 (2016)
61. F. Hardy, A.E. Böhmer, L. de' Medici, M. Capone, G. Giovannetti, R. Eder, L. Wang, M. He, T. Wolf, P. Schweiss, R. Heid, A. Herbig, P. Adelmann, R.A. Fisher, C. Meingast, (2016), arXiv:1605.05485
62. R. Yu, Q. Si, Phys. Rev. Lett. **110**, 146402 (2013)
63. M. Yi, D.H. Lu, R. Yu, S.C. Riggs, J.H. Chu, B. Lv, Z.K. Liu, M. Lu, Y.T. Cui, M. Hashimoto, S.K. Mo, Z. Hussain, C.W. Chu, I.R. Fisher, Q. Si, Z.X. Shen, Phys. Rev. Lett. **110**, 067003 (2013)
64. M. Yi, Z.K. Liu, Y. Zhang, R. Yu, J.X. Zhu, J.J. Lee, R.G. Moore, F.T. Schmitt, W. Li, S.C. Riggs, J.H. Chu, B. Lv, J. Hu, M. Hashimoto, S.K. Mo, Z. Hussain, Z.Q. Mao, C.W. Chu, I.R. Fisher, Q. Si, Z.X. Shen, D.H. Lu, Nat. Commun. **6**, 7777 (2015)
65. G. Giovannetti, de' Medici L, Aichhorn M, Capone M., Phys. Rev. B **91**, 085124 (2015)
66. M.T. Czyzyk, G.A. Sawatzky, Phys. Rev. B **49**, 14211–14228 (1994)
67. M. Karolak, G. Ulm, T. Wehling, V. Mazurenko, A. Poteryaev, A. Lichtenstein, Journal of Electron Spectroscopy and Related Phenomena 181, 11–15 (2010). Proceedings of International Workshop on Strong Correlations and Angle-Resolved Photoemission Spectroscopy 2009
68. H. Park, A.J. Millis, C.A. Marianetti, Phys. Rev. B **90**, 235103 (2014)
69. K. Haule, Phys. Rev. Lett. **115**, 196403 (2015)
70. T. Miyake, K. Nakamura, R. Arita, M. Imada, J. Phys. Soc. Jpn. **79**, 044705 (2010)
71. J. An, A.S. Sefat, D.J. Singh, M.H. Du, Phys. Rev. B **79**, 075120 (2009)
72. D.E. McNally, S. Zellman, Z.P. Yin, K.W. Post, H. He, K. Hao, G. Kotliar, D. Basov, C.C. Homes, M.C. Aronson, Phys. Rev. B **92**, 115142 (2015)
73. M. Zingl, E. Assmann, P. Seth, I. Krivenko, M. Aichhorn, (2016), arXiv:1603.02115
74. S. Lafuerza, H. Gretarsson, F. Hardy, T. Wolf, C. Meingast, G. Giovannetti, M. Capone, A.S. Sefat, Y.J. Kim, P G, L. de' Medici, (2016), arXiv:1607.07417
75. H. Ishida, A. Liebsch, Phys. Rev. B **81**, 054513 (2010)

76. Z.P. Yin, K. Haule, G. Kotliar, Nat. Mat. **10**, 932 (2011)
77. P. Werner, E. Gull, M. Troyer, A. Millis, Phys. Rev. Lett. **101**, 166405 (2008)
78. A. Liebsch, H. Ishida, Phys. Rev. B **82**, 155106 (2010)
79. P. Werner, M. Casula, T. Miyake, F. Aryasetiawan, A.J. Millis, S. Biermann, Nat. Phys. **8**, 331–337 (2012)
80. K. Haule, G. Kotliar, New J. Phys. **11**, 025021 (2009)
81. A. Shorikov, M. Korotin, S. Streltsov, S. Skornyakov, D. Korotin, V. Anisimov, J. Exp. Theor. Phys. **108**, 121–125 (2009). doi:10.1134/S1063776109010154
82. M.S. Laad, L. Craco, (2009), arXiv:0903.3732
83. L. Craco, S. Leoni, EPL (Europhys. Lett.) **92**, 67003 (2010)
84. M. Aichhorn, S. Biermann, T. Miyake, A. Georges, M. Imada, Phys. Rev. B **82**, 064504 (2010)
85. Z.P. Yin, K. Haule, G. Kotliar, Nat. Mater. **10**, 932–935 (2011)
86. T. Misawa, K. Nakamura, M. Imada, Phys. Rev. Lett. **108**, 177007 (2012)
87. T. Misawa, M. Imada, Nat. Commun. **5**, 5738 (2014)
88. E. Bascones, B. Valenzuela, M.J. Calderón, Phys. Rev. B **86**, 174508 (2012)
89. H. Ikeda, R. Arita, J. Kuneš, Phys. Rev. B **82**, 024508 (2010)
90. N. Lanatà, H.U.R. Strand, G. Giovannetti, B. Hellsing, de' Medici L, Capone M., Phys. Rev. B **87**, 045122 (2013)
91. M.J. Calderón, Ld Medici, B. Valenzuela, E. Bascones, Phys. Rev. B **90**, 115128 (2014)
92. D.C. Johnston, Adv. Phys. **59**, 803–1061 (2010)
93. G.R. Stewart, Rev. Mod. Phys. **83**, 1589–1652 (2011)
94. A.V. Chubukov, P.J. Hirschfeld, Phys. Today **68**, 46 (2015)
95. D.P. Johnson, G. Xu, W.G. Yin (eds.), *Iron-Based Superconductivity* (Springer International Publishing, Berlin, 2015)
96. H. Alloul, A. Cano, *Comptes Rendus Physique* 17, 1–4 (2016). Iron-based superconductors / Supraconducteurs à base de fer
97. N.F. Mott, Rev. Mod. Phys. **40**, 677–683 (1968)
98. D. Vollhardt, Rev. Mod. Phys. **56**, 99–120 (1984)

Index

F. Mancini and R. Citro (eds.), *The Iron Pnictide Superconductors*,
Springer Series in Solid-State Sciences 186, DOI 10.1007/978-3-319-56117-2

MIX
Papier aus verantwortungsvollen Quellen
Paper from responsible sources
FSC® C105338

If you have any concerns about our products,
you can contact us on
ProductSafety@springernature.com

In case Publisher is established outside the EU,
the EU authorized representative is:
Springer Nature Customer Service Center GmbH
Europaplatz 3, 69115 Heidelberg, Germany

Printed by Libri Plureos GmbH
in Hamburg, Germany